7 Day

Disease
Resistance
in Plants

Second Edition

Disease Resistance in Plants

Second Edition

J. E. VANDERPLANK

Plant Protection Research Institute
Pretoria, Republic of South Africa

1984

ACADEMIC PRESS, INC.

(*Harcourt Brace Jovanovich, Publishers*)

Orlando   San Diego   San Francisco   New York   London
Toronto   Montreal   Sydney   Tokyo   São Paulo

ACADEMIC PRESS, INC.
Orlando, Florida 32887

United Kingdom Edition published by
ACADEMIC PRESS, INC. (LONDON) LTD.
24/28 Oval Road, London NW1 7DX

Library of Congress Cataloging in Publication Data

Vanderplank, J. E.
    Disease resistance in plants.

    Bibliography: p.
    Includes index.
    1. Plants--Disease and pest resistance. I. Title.
SB731.V25   1984   632'.3   83-21328
ISBN 0-12-711442-4 (alk. paper)

PRINTED IN THE UNITED STATES OF AMERICA

84 85 86 87   9 8 7 6 5 4 3 2 1

Contents

Preface   xi
Preface to the First Edition   xiii

1 Introduction

    Text   1

2 Unspecific Resistance

2.1 Introduction   5
2.2 Diagonal Check for Specificity
     in a Gene-for-Gene Relation   5
2.3 The Resistance Gene Paradox   7
2.4 The Potato–*Phytophthora* System   7
2.5 Hosts and Nonhosts   9
2.6 Host-Specific Toxins   11
2.7 Discussion   12

3 Host Plants: Phenotypic Variation and Gene Numbers

3.1 Introduction   13
3.2 Partition of Variance   14

3.3   Discontinuous Variation with Many Genes Involved:
       Pseudomonogenic Resistance   16
3.4   Discontinuous Variation with Few Genes Involved   19
3.5   Discontinuous Variation and Cytoplasmic Inheritance   20
3.6   Continuous Variation   21
3.7   Background to the Polygene Story   24
3.8   The Error of Expecting Safety in Numbers:
       Additive Variance   25
3.9   Experimental Difficulties in Partitioning Variance   26

4   The Pathogen: Epistasis and Virulence

4.1   Introduction   28
4.2   The ABC–XYZ Classification
       and Diallel Gene Pairing   29
4.3   Virulence Dissociation   31
4.4   Epistasis–Environment Interaction   33
4.5   Danger in Artifacts   34
4.6   Virulence Association   35
4.7   Definition of a Physiological Race   37

5   Adaptation of the Pathogen to the Host: Wheat Stem Rust in Australia

5.1   Introduction   39
5.2   Direct Adaptation of the Pathogen to the Host   40
5.3   Indirect Adaptation of the Pathogen to the Host   43
5.4   Discussion   47

6   Mutation in the Pathogen from Avirulence to Virulence

6.1   Variable Mutation Rates   49
6.2   Wild-Type Pathogen Populations   50
6.3   Viral Diseases   51
6.4   Bacterial Diseases   52

6.5   Fungal Diseases   53
6.6   Inoperative and Operative Mutation   55
6.7   Epidemiological Mutation   55

7   Horizontal and Vertical Resistance

7.1   Definitions in a Two-Variable System   57
7.2   The Geometric Illustration   58
7.3   Illustration by Analysis of Variance   60
7.4   Illustration by Ranking Order   61
7.5   Vertical Resistance Effective Only against Initial Inoculum 64
7.6   Interrupted or Uninterrupted Presence
      of Disease or Inoculum   66
7.7   Vertical Partial Resistance   71
7.8   Adult-Plant Resistance: Scheibe's Rule   72
7.9   Higher-Order Interactions   74
7.10 How Realistic Is the Definition
      of Horizontal Resistance?   75
7.11 A Third Variable   76
7.12 Qualitative and Quantitative Variation
      in Host and Pathogen   77
7.13 Unclassified Resistance   79
7.14 Pseudospecificity   79
7.15 Remnants of Horizontal Resistance   80
7.16 Horizontal Resistance and Stabilizing Selection   80

8   Remnants of Resistance

8.1   Three Questions   82
8.2   The Vertifolia Effect   83
8.3   Ghost Resistance   88
8.4   The Either/Or Avirulence/Virulence Error   89
8.5   The Horizontal Resistance Equivalent   91

9   Protein Polymorphism and Vertical Resistance

9.1   Introduction   92
9.2   Molecular Association in Specific Susceptibility   93

9.3   Molecular Storage of Massive Variation   93
9.4   Endothermic Susceptibility   94
9.5   Individuality in Temperature Responses   94
9.6   Vertical Resistance   95
9.7   Coping with Recessive Resistance   95
9.8   Test to Distinguish Vertical from Horizontal
       Partial Resistance   95
9.9   The Gene-for-Gene Hypothesis   96

10 Genes for Susceptibility

10.1   Introduction   99
10.2   Vavilov's Rule   100
10.3   The Corollary of Vavilov's Rule   100
10.4   Some Misinterpreted Evidence   101
10.5   Biotropic Semibenign Infection   101
10.6   Wounds and Infections: Role of Peroxidase   102
10.7   Ultrastructural Evidence   104
10.8   Protein Polymerization   104
10.9   The Pathogen's Protein   105
10.10 Theory of Reciprocal Mutation   106

11 Sink-Induced Loss of Resistance

11.1   Introduction   107
11.2   Stalk Rot of Maize   108
11.3   Sugar and Resistance to Maize Stalk Rot   111
11.4   Ecological Topics   112
11.5   The Vertifolia Effect Again   113
11.6   Low-Sugar Disease Processes   113
11.7   Theories about High-Sugar Resistance   115
11.8   Discussion   116

12 High-Sugar Disease Processes and Biotrophy

12.1   Introduction   117
12.2   High-Sugar Susceptibility   118

12.3   The Sugar Effect   119
12.4   Reversal of Resistance to Powdery Mildew   120

13 Epidemiological Effects of Vertical Resistance

13.1   Resistance in Relation to an Increase
        in the Population of the Pathogen   122
13.2   The Effect of Vertical Resistance:
        The General Rule   123
13.3   The Effect of Vertical Resistance:
        Some Illustrative Data   125
13.4   Vertical Resistance: The Price of Varietal Popularity   128
13.5   Vertical Resistance: The Enhancing Effect
        of Horizontal Resistance   129
13.6   Generalized Disease Progress Curves
        for the Study of the Effects of Resistance   131
13.7   The Compound Interest Equation:
        Logarithmic Increase of Disease   133
13.8   An Equation for the Effect of Vertical Resistance   135
13.9   An Analysis of Some Experimental Data   137
13.10   Graphical Representation of Equation (13.1)   138
13.11   Vertical Resistance: The Quantitative Effect
        of Varietal Popularity   139
13.12   The Effect of Vertical Resistance
        after the Logarithmic Phase of the Epidemic   141
13.13   Independence of Initial Inoculum
        and the Logarithmic Infection Rate:
        The Start of an Epidemic   142
13.14   Appendix: Vertical Resistance That Reduces
        the Infection Rate   143

14 Epidemiological Effects of Horizontal Resistance

14.1   Introduction   146
14.2   History of Blight Resistance
        in Three Potato Varieties   147

14.3   Horizontal Resistance and a Reduced Infection Rate   149
14.4   Components of Horizontal Resistance   150
14.5   Ontogenic Effects   153

15 Slow Rusting of Cereal Crops

15.1   Introduction   154
15.2   Late Rusting and Slow Rusting   156
15.3   Maize Rust   156
15.4   Resistance before and Susceptibility after Flowering   158
15.5   Sink-Associated and Sink-Induced Loss of Resistance   159
15.6   Breeding for Horizontal Slow-Rusting Resistance   160

16 Resistance against Endemic Disease

16.1   Introduction   162
16.2   The Progeny/Parent Ratio   164
16.3   Infection Rates and Disease Levels   165
16.4   The Progeny/Parent Ratio and Latent Period
        in the Strategy of Using Disease Resistance   166
16.5   Horizontal and Vertical Resistance in the Strategy
        of Using Disease Resistance   167
16.6   Endemic Disease in the Tropics   167
16.7   Appendix   168

17 Heterogeneous Host Populations
       and the Accumulation of Resistance Genes

17.1   Introduction   171
17.2   Mixed Varieties and Multilines   172
17.3   Heterogeneity versus Gene Accumulation   173
17.4   Essential Purpose of Mixed Varieties and Multilines   174
17.5   Role of the ABC–XYZ Groups:
        Stabilizing Selection versus Heterogeneity   175

Bibliography   177

Index   191

Preface

Most of the chapters in this book include information not available when the first edition appeared in 1968. Its comprehensive coverage will be of interest to plant pathologists and plant breeders. Both are concerned with developing new cultivars possessing genetic resistance to diseases; both contribute special skills. But skills alone are not enough. To be properly employed relevant genetic, epidemiologic, biochemical, and biometric principles must be understood. The aim of this work is to help provide such an understanding.

It is known that an incompatibility can exist between high resistance and high yields of grain, fruits, tubers, and bolls. These are metabolic sinks that sometimes induce a loss of resistance. Evidence indicates that the loss, when it occurs, is in horizontal resistance and that vertical resistance escapes sink-induced losses. The desire to obtain the greatest possible yields is perhaps a reason why plant breeders prefer vertical resistance if it is available and can be stabilized by stabilizing the selection of the pathogen. In conjunction with lost resistance, sugar seems to be the most important substance drained by the sink, which suggests that to maintain high yields the best forms of resistance are those enhanced by low sugar content. They are more likely to be found against biotrophy than necrotrophy. This book includes discussions on stabilizing selection, sugar, biotrophy, and necrotrophy.

This publication coincides with a burst of activity in plant genetic engineering, which provides new methods for manipulating genetic material. It will aid plant pathologists and plant breeders to discover what genetic material they wish to manipulate. The nature of resistance and resistance genes is investigated using information not analyzed elsewhere and with results that are a necessary prerequisite for the breeding of resistant varieties by genetic engineering.

Breeding for disease resistance involves two organisms: the host plant and the pathogen. The breeder changes the host; in doing so he may also change the pathogen if host and pathogen are genetically interlocked as they are in a gene-for-gene system. The pathogen changes by reflection from the host. The plant breeder can put genetic reflection to use in order to weaken the pathogen and reduce disease. Reflective genetic engineering of the pathogen may yet become an important part of plant breeding for disease control. A major portion of one chapter is devoted to a discussion of the effect of reflected virulence on the structure of the pathogenic population.

Agronomists and horticulturists must know the disease resistance of the varieties they recommend, and disease resistance is one of the features that determines acceptability to farmers. This book, by comprehensively covering disease resistance and stressing its limitations as well as its advantages, helps provide a suitable basis for making the proper agronomic and horticultural recommendations.

I thank Drs. N. H. Luig and R. A. McIntosh for sending me recent reports on wheat stem rust in Australia.

J. E. Vanderplank

Preface to the
First Edition

The purpose of this book is to inquire into the role of resistance in plant disease. It discusses the nature of resistance and how it can best be used to protect crops from disease.

The chapters contain much that is new. First, it is shown that there are two types of pathogenic races and of pathogenicity which correspond to the two types of resistance. The evidence for this seems clear and incontrovertible. Second, the strength of genes for vertical resistance in host plants is measured. This is done by introducing the concept of relative half-lives of matching pathogenic races. The argument is simple and the evidence direct. Third, in discussing vertical resistance to obligate parasites, host–host–pathogen systems replace the host–pathogen systems ordinarily considered in current literature. This conforms with the self-evident fact that there must be two or more host genotypes within the epidemic area if there is to be satisfactory vertical resistance to obligate parasites that do not live long outside their host plants. The change to host–host–pathogen systems will be found to clarify many practical problems in the use of resistance. Fourth, evidence is presented that in its effects on disease a change in the polygenic horizontal resistance of the host plants is often identical with a change in the aggressiveness of the pathogen or of that in the environment. It seems that the genes of horizontal resistance are often not special resistance genes, but ones concerned in the normal metabolic processes of healthy plants. For this reason there may be large untapped reserves of horizontal resistance in many crops. Fifth, a theory is given of vertical and horizontal resistance: Resistance is vertical if to overcome it the pathogen must become less aggressive on susceptible varieties of the host; it is horizontal if to overcome it the pathogen must become more aggressive on susceptible varieties as well. The theory fits the known facts.

Two topics introduced in my previous book ("Plant Diseases: Epidemics and Control," Academic Press, 1963) now receive more attention. The effect of resistance on the progress of disease in the field has a chapter to itself; disease progress curves are used to illustrate the effect. Stabilizing selection is the main topic of two chapters. It is at the core of stable vertical resistance conferred by strong genes. In emphasizing stabilizing selection the book departs from the conventional treatment of pathogenic races. Instead of an emphasis on how races arise, the theme of countless papers in the literature, it places the emphasis on how fit races are to survive after they have arisen, a theme hitherto much neglected.

The theory of multilines is examined, particularly in relation to stabilizing selection. The great unanswered question is whether there are enough strong genes to maintain adequate resistance in the long run.

The work is intended for plant pathologists, because sooner or later they must deal with matters of resistance; for plant breeders, because breeding for resistance is one of the important reasons for creating new varieties; and for those interested in ecology, because nowhere is natural balance better illustrated than by the interaction of plant host, pathogen, and environment.

The chapters interlock because resistance theories are becoming consistently integrated. However, without too much repetition, I tried to make each chapter an independent unit to reduce the need for tedious cross references.

Pretoria, South Africa   *J. E. Van der Plank*
June 1968

1

Introduction

Our daily bread comes from wheat fields protected against disease by genetic resistance; chemical protection is practically limited to seed dressings. Our meat, dairy products, and eggs come ultimately from maize, soybeans, pastures, and fodder crops protected against disease by genetic resistance. Our food is sweetened by sugar; a change from susceptible to resistant cultivars saved the sugarcane industry almost worldwide from mosaic disease, and the sugar beet industry in areas west of the Rocky Mountains from curly top disease. So the story continues, with genetic resistance being seen as a pillar of agriculture.

Disease resistance has come from many sources. It has come from transgressive segregation and from other varieties within the crop species. Examples are innumerable. It has come from different but related species and genera. Already at the beginning of the century Orton had transferred to watermelons (*Citrullus lanatus*) the resistance of stock citrons (*C. vulgaris*) to *Fusarium oxysporum* f. sp. *niveum*. Modern tomato cultivars have gained disease resistance from other *Lycopersicon* spp. Hexaploid bread wheat has been fortified by genes from tetraploid and diploid *Triticum*, and intergenerically from *Secale*, *Agropyron*, and *Aegilops*. Knott and Dvořák (1976) have reviewed the use of alien germ plasm for developing disease-resistant cultivars. The advantage of using alien germ plasm is great; an almost limitless source of resistance is tapped. The difficulties are equally great. Fertilization is often difficult; crosses fail soon after fertilization; and hybrid plants may die before flowering or may be sterile. The interspecific and intergeneric

1

transfer of a gene for disease resistance involves the transfer of a chromosome segment from a donor species or genus to a recipient. The substitution of an alien segment may cause undesirable duplications or deletions, and the segment is likely to carry unwanted genes linked to the wanted resistance gene. Getting rid of the unwanted genes requires crossing-over, and this becomes more difficult when the donor is foreign and the chromosomes are not homologous but only homeologous. Nevertheless, despite the difficulties, there is a large body of literature of successful and useful donations by alien species and genera.

Natural mutation to resistance occurs. In potatoes, mutation from smooth-skinned to russet-skinned tubers commonly brings with it increased resistance to scab caused by *Streptomyces scabies*. Mutagens have been used to introduce resistance, as in peppermint (*Mentha piperita*) to *Verticillium albo-atrum* (Murray, 1969; Todd *et al.*, 1977). Stolons of the susceptible peppermint cultivar Mitcham were treated with neutrons or X-rays to give mutants both resistant and of high horticultural quality. Other examples are quoted in a review of mutagenesis by Simons (1979).

Protoclones are the newest way of developing disease-resistant variants. Shepard *et al.* (1980) found increased resistance to the blights caused by *Phytophthora infestans* and *Alternaria solani* in clonal populations regenerated from mesophyll cell protoplasts of the potato cultivar Russet Burbank. They solved the technical problems of regenerating plants from mesophyll protoplasts and then compared a number of "protoclones," as they called them. Some were more resistant than others to *P. infestans*, and some to *A. solani*. It is still too early to state whether these resistant protoclones differ essentially from the meristem mutants known as bolters and semibolters, which often have increased resistance to blights. But, be the resistant protoclones horticulturally successful or not, the work of Shepard *et al.* marks the entry of genetic engineering into plant breeding for disease resistance. A new era has begun.

Genetic engineering is a general term used to describe cellular and molecular methods for altering the genetics of organisms. Research has developed rapidly in two directions. In the direction of tissue and cell culture, plants (and other multicellular organisms) can be reduced to single cells grown under aseptic laboratory conditions. In the direction of molecular biology, the emphasis is on recombinant DNA techniques. Single genes isolated from one organism can be introduced into another, thereby transforming the recipient genetically. Chromosomes, chromosome segments, DNA preparations, and cellular organelles can also be transferred to a recipient.

Cell cultures can be started from plant tissue by inoculation into an appropriate medium and maintained aseptically. A callus of rapidly dividing cells develops and can be maintained by repeated transfers. Callus cells inoculated into a liquid medium with continuous agitation form a suspension culture of single cells and small cell aggregates. Protoplasts or plant cells without walls are produced by treatment with enzymes. Protoplasts have unique properties. They fuse with other protoplasts or absorb foreign genes, chromosomes, and organelles. From manipulated protoplasts whole plants can be regenerated by appropriate techniques.

At present these methods are far from routine. They work well with some species, notably tobacco, but there are unsolved difficulties with other plants such as cereals and soybeans. Where they do work, new procedures for producing disease-resistant plants are available. Protoplasts unlock variation, as the results of Shepard *et al.* (1980) show. When disease is caused by a toxin, millions of protoplasts can be screened for resistance in a small flask; they are the equivalent of thousands of acres of growing plants. Protoplast fusion or *in vitro* pollination (fertilization of placentas) can produce hybrids between species and genera of plants where normal sexual procedures fail; this increases the range of donors of resistance. Recombinant DNA methods coupled with protoplast-to-plant regeneration can be aimed at introducing single genes for resistance uncluttered by association with deleterious genes.

Techniques of genetic engineering will develop swiftly. Our particular concern is with how to use them.

Specificity resides in susceptibility. Resistance is unspecific; this is the topic of Chapter 2. Unspecific resistance is the single most fortunate fact in the whole of plant pathology. It makes susceptibility the exception, not the rule. For genetic engineers it makes the whole plant world a treasure-house of resistance genes that we call nonhost resistance genes. Putting these genes to work is an essential part of genetic engineering.

Wheat is resistant to all the rust fungi that attack maize; maize is resistant to all the rust fungi that attack wheat. Wheat is a nonhost of the maize rust fungi, and maize a nonhost of the wheat rust fungi. One of the tasks of coming years is to get rustfree wheat and maize by a selective exchange of nucleotide base sequences. But first we must learn what the relevant sequences are. The available evidence suggests that there are at least three. First, there is (the evidence suggests) a variable sequence that determines specificity in gene-for-gene systems. Second, there is a conserved sequence that seems to be involved in processes that cause the pathogen to be turned on to produce an elicitor even in some nonhost plants. Third, there are sequences that determine the tertiary structure of the coded protein and the quaternary structure of protein polymers and copolymers.

In gene-for-gene systems the variable sequence in the host determines a variable sequence in the pathogen, and a pair of resistance genes in the host determines a pair of sequences in the pathogen. This enables the host to force unfavorable epistatic interactions on the pathogen; these interactions are the topic of Chapter 4. Genetic engineering of the host can by reflection become genetic engineering of the pathogen, to the pathogen's detriment. Reflected genetic engineering of the pathogen is perhaps possible not only in gene-for-gene systems but also in all systems in which host and pathogen have interlocking genetic arrangements. As Chapter 4 shows, there are strong environmental effects to be considered. Chapters 9 and 10 probe the nature of resistance/susceptibility genes in the host and their relation to virulence/avirulence genes in the pathogen.

Great advances in genetic engineering are for the future. For the present the greatest need is the breeding of resistant varieties in tropical agriculture by techniques immediately available. Plant disease is rife in the tropics. Wellman (1972) estimated that of diseases causing significant losses in the Americas the ratio is about 10 tropical for every temperate-zone disease. For economic and environmental reasons, disease control must be mainly by resistant cultivars. Much has been and is being done, and much success has been achieved. Tropical disease has some special features, such as a greater continuity of inoculum than in temperate countries and a greater proportion of endemic as distinct from epidemic disease. This affects the choice of the sort of resistance to be used, and it has been thought fit to include a special chapter (Chapter 16) about endemic disease and related topics.

Resistance and susceptibility are opposites; a more resistant plant is a less susceptible plant. Chapter 2 establishes the fact, crucial for molecular and physiological plant pathology, that host–pathogen specificity is determined by susceptibility. Thereafter, the chapters proceed from topics mainly genetic to topics mainly epidemiological, with some biometric and biochemical discussion between. The choice of pathogens for illustration is almost automatic; those pathogens figure most about which there is most information in the literature. They are also generally the most destructive pathogens, and therefore the most usefully included for discussion.

2
Unspecific Resistance

2.1 INTRODUCTION

There is specificity in all plant disease. Disease is not a random pairing among thousands of host plants and thousands of pathogens. It is selective, occurring in particular pairs. Wheat and *Puccinia graminis* pair to cause wheat stem rust. Even the fungus *Phymatotrichum omnivorum*, despite its name, has preferences. It preferentially attacks dicotyledons.

Where does this specificity lie? Is it in susceptibility, or is it in resistance? The answer is clear. What specificity there is, is in susceptibility. Resistance is wholly unspecific.

2.2 DIAGONAL CHECK FOR SPECIFICITY IN A GENE-FOR-GENE RELATION

Flor's (1942) gene-for-gene hypothesis will be discussed in Chapter 9. Here we merely state it: For every gene for resistance in the host plant there is a corresponding and specific gene for virulence in the pathogen.

Table 2.1 illustrates the hypothesis. In general terms, a plant with the resistance gene *R1* is, by hypothesis, susceptible to a pathogen with the

TABLE 2.1

The Diagonal Check for Specificity in a Gene-for-Gene Relation[a]

	Plant				
Pathogen	R1R1[b]	R2R2	R3R3	R4R4	R5R5
v1v1	Susceptible[c]	Resistant	Resistant	Resistant	Resistant
v2v2	Resistant	Susceptible	Resistant	Resistant	Resistant
v3v3	Resistant	Resistant	Susceptible	Resistant	Resistant
v4v4	Resistant	Resistant	Resistant	Susceptible	Resistant
v5v5	Resistant	Resistant	Resistant	Resistant	Susceptible

[a] Plant reaction when resistance genes R1, R2, R3, R4, and R5 at five loci interact with virulence genes v1, v2, v3, v4, and v5 at five loci in the pathogen.

[b] Resistance is assumed to be dominant, and RR can be replaced by Rr. Virulence is assumed to be recessive. This is usual but not invariable or necessary for the illustration; recessive resistance and dominant virulence are known.

[c] The plant is susceptible.

corresponding virulence gene v1 and not to a pathogen without this particular virulence gene irrespective of how many other virulence genes it may have. Table 2.1 illustrates the hypothesis for five host varieties, each carrying different resistance genes, and five isolates of the pathogen, each carrying different virulence genes.

There are 25 entries in Table 2.1. Of these, 5 are on a diagonal, and these 5 are for susceptible plants; 20 are off the diagonal, and these 20 are for resistant plants. The diagonal is where host and pathogen correspond and represents a specific relationship. Off the diagonal there is no specific relationship. A plant with the resistance gene R1 is susceptible only, i.e., specifically, to a pathogen possessing the virulence gene v1 with which the gene R1 corresponds diagonally. It is resistant unspecifically to pathogens possessing any or all of the virulence genes v2, v3, v4, and v5 with which the gene R1 does not correspond diagonally. So too a plant with the resistance gene R2 is susceptible only, i.e., specifically, to a pathogen possessing the virulence gene v2 and is unspecifically resistant to pathogens with any or all of the virulence genes v1, v3, v4, and v5, and so on. All the specificity embodied in the gene-for-gene hypothesis is in susceptibility. Resistance is unspecific.

Table 2.1 can be expanded indefinitely. Browder (1980) recognized 35 genes in wheat for resistance to *Puccinia recondita*. If one accepts that they are all in gene-for-gene systems, they would need a corresponding 35×35 table, in which 35 entries would be on the diagonal, indicating specificity, and $34 \times 35 = 1190$ off the diagonal, indicating an unspecific relation.

2.3 THE RESISTANCE GENE PARADOX

Resistance genes are identified individually, i.e., specifically, but the resistance they give is unspecific. That is the paradox illustrated by Table 2.1. The resistance genes *R1*, *R2*, . . . are identified and numbered individually; they are specific genes. But in relation to the *v* genes, i.e., in relation to whether disease will or will not occur, resistance is unspecific. This is the inescapable logic of the diagonal (or more precisely the off-diagonal) arrangement. Specific resistance genes and unspecific resistance go together.

The diagonal check explains the paradox. A resistance gene in the host is identified by the virulence allele, *not the avirulence allele*, in the corresponding locus in the pathogen. That is, it is identified in a state of susceptibility, not resistance.

An erroneous belief in specific resistance, as distinct from specific resistance genes, has led to much "barking up the wrong tree." Among the examples of misdirected "barking" are the phytoalexin and hypersensitivity theories of host–pathogen specificity.

2.4 THE POTATO–*Phytophthora* SYSTEM

A hypothetical gene-for-gene relation has emerged in a different guise in a system for identifying pathogenic races. The system was first devised by Black *et al.* (1953) for races of *Phytophthora infestans* and resistance genes in potato, but has since been applied in various forms to other diseases.

The system is illustrated by Table 2.2. There are 4 genes for resistance included in the table (out of the 11 or more known in *Solanum*) and a corresponding $2^4 = 16$ races of *P. infestans*. The genes are numbered *R1*, *R2*, *R3*, and *R4* and are entered singly and in combination. A plant with the gene *R1* is susceptible to race (1) or any other race with 1 in its designation, but resistant to races (2), (3), and (4) or races with 2, 3, or 4 in any combination in their designation. So too a plant with the gene *R2* is susceptible to race (2) or any other race with the number 2 in it, but is resistant to races (1), (3), and (4). A plant with the two genes *R1* and *R2* is susceptible to race (1, 2) or any other race with both the numbers 1 and 2 in it, but is resistant to all other races. And so on.

The more the resistance genes the clearer it becomes that susceptibility is specific and resistance unspecific. In a 4-gene system a plant with the genes *R1*, *R2*, *R3*, and *R4* combined is susceptible only, i.e., specifically, to race (1, 2, 3, 4). It is resistant to any or all of the other 15 races, i.e., resistance is unspecific. In a 10-gene system (with $2^{10} = 1024$ races) a plant

TABLE 2.2

International System of Designating Interrelationships of Genes of Potatoes and Races of *Phytophthora infestans*[a]

Gene(s)	Race															
	(0)	(1)	(2)	(3)	(4)	(1, 2)	(1, 3)	(1, 4)	(2, 3)	(2, 4)	(3, 4)	(1, 2, 3)	(1, 2, 4)	(1, 3, 4)	(2, 3, 4)	(1, 2, 3, 4)
r	S	S	S	S	S	S	S	S	S	S	S	S	S	S	S	S
R1	R	S	R	R	R	S	S	S	R	R	R	S	S	S	R	S
R2	R	R	S	R	R	S	R	R	S	S	R	S	S	R	S	S
R3	R	R	R	S	R	R	S	R	S	R	S	S	R	S	S	S
R4	R	R	R	R	S	R	R	S	R	S	S	R	S	S	S	S
R1, R2	R	R	R	R	R	S	R	R	R	R	R	S	S	R	R	S
R1, R3	R	R	R	R	R	R	S	R	R	R	R	S	R	S	R	S
R1, R4	R	R	R	R	R	R	R	S	R	R	R	R	S	S	R	S
R2, R3	R	R	R	R	R	R	R	R	S	R	R	S	R	R	S	S
R2, R4	R	R	R	R	R	R	R	R	R	S	R	R	S	R	S	S
R3, R4	R	R	R	R	R	R	R	R	R	R	S	R	R	S	S	S
R1, R2, R3	R	R	R	R	R	R	R	R	R	R	R	S	R	R	R	S
R1, R2, R4	R	R	R	R	R	R	R	R	R	R	R	R	S	R	R	S
R1, R3, R4	R	R	R	R	R	R	R	R	R	R	R	R	R	S	R	S
R2, R3, R4	R	R	R	R	R	R	R	R	R	R	R	R	R	R	S	S
R1, R2, R3, R4	R	R	R	R	R	R	R	R	R	R	R	R	R	R	R	S

[a] S means the potato plant is suceptible; R, it is resistant.

with the 10 genes, *R1* to *R10*, is susceptible only, i.e. specifically, to race (1, 2, 3, 4, 5, 6, 7, 8, 9, 10). It is resistant to any or all of the other 1023 races, i.e., resistance is unspecific.

There is race-specific susceptibility but not race-specific resistance. Nevertheless, the term race-specific resistance is commonly used in the literature. It is both false and clumsy, and should be shunned.

A diagonal check occurs at three different levels in Table 2.2 It occurs when the four single genes *R1*, *R2*, *R3*, and *R4* interact with the four simple races (1), (2), (3), and (4). It occurs when the six pairs of genes *R1* and *R2*, *R1* and *R3*, *R1* and *R4*, *R2* and *R3*, *R2* and *R4*, and *R3* and *R4* interact with the corresponding six races (1, 2), (1, 3), (1, 4), (2, 3), (2, 4), and (3, 4). It occurs when the four groups of three genes *R1*, *R2*, and *R3*; *R1*, *R2*, and *R4*; *R1*, *R3*, and *R4*; and *R2*, *R3*, and *R4* interact with the corresponding four races (1, 2, 3), (1, 2, 4), (1, 3, 4), and (2, 3, 4). At each level of resistance there is hypothetically a corresponding level of virulence, complexity matching complexity. At each level susceptibility is on the diagonal and therefore specific, and resistance is off the diagonal and therefore unspecific.

2.5 HOSTS AND NONHOSTS

The previous two sections dealt with susceptibility and resistance within narrow host taxa. Now we widen the scope.

Table 2.3, comparable with Table 2.1, deals with the rust diseases of three grasses, two legumes, and two other dicotyledons. Again, the diagonal represents correspondence and specificity between host and pathogen; and susceptibility, not resistance, is on the diagonal.

Maize is susceptible specifically to *Puccinia sorghi*, *P. polysora*, and *Physopella zeae*. It is host to them. Maize is resistant unspecifically not only to the other 10 rust fungi listed in the table but also to every other of the more than 1000 rust fungi known to mycologists. It is a nonhost to them. *Mutantis mutandis*, the same applies to the other plants in the table. Because of the wider taxa involved, specificity is not so fine as in the previous two sections. *Puccinia graminis*, *P. recondita*, and *P. striiformis* all attack both wheat and barley. But what specificity there is, is in susceptibility.

What holds for the rust fungi holds for many other groups. Table 2.4 refers to the anthracnose fungi, without changing the conclusions. Resistance is off the diagonal and therefore unspecific, which is the point of this chapter.

TABLE 2.3

Diagonal Plant–Pathogen Relations in Some Rust Diseases

Pathogen	Plant						
	Maize	Wheat	Barley	Potato	Cotton	Soybean	Alfalfa
Puccinia sorghi, *P. polysora,* *Physopella zeae*	*Susceptible*[a]	Resistant[b]	Resistant	Resistant	Resistant	Resistant	Resistant
Puccinia graminis, *P. recondita,* *P. striiformis*	Resistant	*Susceptible*	*Susceptible*	Resistant	Resistant	Resistant	Resistant
Puccinia hordei	Resistant	Resistant	*Susceptible*	Resistant	Resistant	Resistant	Resistant
Puccinia pittieriana, *Aecidium cantensis*	Resistant	Resistant	Resistant	*Susceptible*	Resistant	Resistant	Resistant
Puccinia cacabata, *Phakopsora gossypii*	Resistant	Resistant	Resistant	Resistant	*Susceptible*	Resistant	Resistant
Phakopsora pachyrhizi	Resistant	Resistant	Resistant	Resistant	Resistant	*Susceptible*	Resistant
Uromyces striatus	Resistant	Resistant	Resistant	Resistant	Resistant	Resistant	*Susceptible*

[a] The plant is susceptible, i.e., it is a host.
[b] The plant is resistant (= immune), i.e., it is a nonhost.

TABLE 2.4

Diagonal Plant–Pathogen Relations in Some Anthracnose Diseases

	Plant				
Pathogen	Maize, wheat	Potato	Cotton	Soybean	Alfalfa
Colletotrichum graminicola	Susceptible[a]	Resistant[b]	Resistant	Resistant	Resistant
C. atramentarium	Resistant	Susceptible	Resistant	Resistant	Resistant
Glomerella gossypii	Resistant	Resistant	Susceptible	Resistant	Resistant
C. dematium, G. glycines	Resistant	Resistant	Resistant	Susceptible	Resistant
C. trifolii	Resistant	Resistant	Resistant	Resistant	Susceptible

[a] The plant is susceptible, i.e., it is a host.

[b] The plant is resistant (= immune), i.e., it is a nonhost.

2.6 HOST-SPECIFIC TOXINS

To workers on host-specific toxins must go the credit for being the first to understand clearly and demonstrate specific susceptibility and unspecific resistance. The name they chose, host-specific toxins, embodies this understanding.

The primary molecular interaction between the pathogen's toxin and the host's receptor is one of susceptibility. A precise correlation exists between toxin production by the pathogen and susceptibility of the host. The toxins reproduce the disease. They cause the same changes in the plants as the toxin-producing fungi do. Plants susceptible to the fungus are sensitive to its toxin. Plants susceptible to the fungus and sensitive to its toxin have been crossed with plants resistant to the fungus and insensitive to its toxin. The precise correlation is carried through to the segregation progeny. Progeny susceptible to the fungus are sensitive to its toxin, and those resistant to the fungus are insensitive to its toxin.

There is no specificity in resistance. The toxins are unspecifically inactive in all plants, including all nonhosts, that are resistant to the fungi that produce the toxins.

Of the diseases known to be caused by host-specific toxins, Victoria blight of oats caused by *Helminthosporium victoriae*, southern leaf blight of maize caused by *H. maydis* race T, and the leafspot disease of maize caused by *H. carbonum* have been the most closely studied. The literature has been widely reviewed, among others by Scheffer and Yoder (1972), Scheffer (1976), and Yoder (1980, 1981).

2.7   DISCUSSION

The pathogens used for illustration in this chapter range from biotrophs (feeding on living cells) to necrotrophs. The rust fungi are biotrophs. *Phytophthora infestans* is a hemibiotroph; for 3–4 days after infection it feeds biotrophically and then changes to necrotrophy. It is during the biotrophic phase, within a few hours after infection, that specificity is determined. Less is known about the anthracnose fungi except *Colletotrichum lindemuthianum*. This fungus is also a hemibiotroph; for 4 days after infection it causes no visible damage to the host protoplast, after which it changes to necrotrophy, destroying cell walls and cells in the process (Mercer *et al.*, 1974). Again, specificity is determined early, during biotrophy. Finally, pathogens with host-specific toxins are necrotrophs, and specificity is determined during necrotrophy.

Over the whole range from biotrophy to necrotrophy, whenever specificity is analyzed, it is found to lie in susceptibility. The degree of specificity may be high, as in most rust diseases or diseases caused by host-specific toxins, or it may be low, as when the pathogen has a wide host range. But whatever specificity there is, whether high or low, resides in susceptibility alone.

This book is about resistance and discusses specificity. It is well to remember that what may seem superficially to be specific resistance is really only the reflection of specific susceptibility. Sometimes it is enough to see resistance by reflection; at other times clearer vision is needed.

3

Host Plants: Phenotypic Variation and Gene Numbers

Variation is the plant breeder's material. From variable material he selects resistant parents with which to make his crosses. Later, he selects resistant lines from the segregating and therefore variable progenies. Basic information about variation helps to decide procedure. In varying populations, how much of the total phenotypic variation is environmental and, being non-heritable, useless for breeding programs? Of the genetic and therefore heritable variation, how much is additive? How many genes contribute to the variation? Is resistance dominant or recessive? Does resistance vary discontinuously or continuously?

Discontinuous variation is also called qualitative variation. With discontinuity, resistance in plants in a population falls into distinct and separate categories. Continuous variation is commonly called quantitative variation. With continuity, resistance varies smoothly without interruption from the most susceptible to the most resistant plants. The distinction between discontinuous and continuous variation is not always clear. There are often intermediates. But the distinction is clear enough, and the consequences of the distinction important enough, for continuity or lack of it to be used as the basis for discussion.

Another classification partitions total phenotypic variance into environmental, nonheritable variance and genetic, heritable variance. In turn, genetic variance is partitioned into additive variance, dominance variance, and epistatic variance, depending on the absence or presence of interaction between alleles in a locus or between genes in different loci.

3.2 PARTITION OF VARIANCE

Environmental variance is the plant breeder's bane. It makes plants or experimental plots vary even when they are genetically uniform and have received the same treatment. It is the source of "escapes," i.e., plants that seem to have resistance but for one reason or another have simply escaped infection. Its special relevance to this chapter is that it tends to smother other variation and make the phenotype vary continuously where the genotype itself varies discontinuously.

Additive variance makes a hybrid take equally after both parents. If all of the genetic variance is additive, a cross will fall midway between its parents. Variance is additive if the substitution of one allele for another produces a shift that is the same irrespective of what other genes are present, i.e., if the effect of change from *aa* to *aA* is the same as the effect of change from *aA* to *AA*, irrespective of genes in other loci. Additive variance is particularly relevant to continuous phenotypic variation.

Dominance makes a hybrid take more after one parent than the other. The effects of change from *aa* to *aA* and from *aA* to *AA* are unequal. Dominance is a major source of discontinuity in phenotypic variation.

Dominance variance applies equally to dominant resistance and recessive resistance, which is dominant susceptibility. There is nothing incongruous about including recessive resistance under dominance variance. Recessive resistance has featured largely in the control of disease, though not as much as dominant resistance. The first Mendelian study of the inheritance of disease resistance was with recessive resistance in wheat to *Puccinia striiformis* (Biffen, 1905). Recessive monogenic resistance stopped the epidemic of Victoria blight in oats caused by *Helminthosporium victoriae*, and has remained effective ever since. The recessive monogene *rhm* effectively controls southern maize leaf blight caused by *H. maydis* (*Cochliobolus heterostrophus*) (Smith, 1975). The literature of disease of the various crop plants frequently mentions recessive resistance.

Dominance reversal is known. Temperature can change dominance to recessivity. The gene *Sr6* in wheat for resistance to *P. graminis tritici* is dominant at temperatures of 18°C or less, but recessive at temperatures above 24°C (Luig and Watson, 1965). In tomato the gene *Tm1* conditioned dominant resistant to 53% of the isolates of tobacco mosaic virus at 17°C but to none at 30°C; the gene *Tm2* conditioned dominant resistance to 80% of the isolates at 17°C but to only 34% at 30°C; and the gene *Tm2a* conditioned dominant resistance to 58% of the isolates at 17°C but to only 7% at 30°C (Cirulli and Ciccarese, 1975; Vanderplank, 1982). Environmental factors other than temperature have been reported. Catherall *et al.* (1970) reported that the gene *Yd2* in barley for tolerance to barley yellow dwarf virus is fully dominant or fully recessive, depending on how fast the barley plants were growing. The constitution of the pathogen may also change resistance in the host from dominant to recessive. Hooker and Saxena (1971) obtained results with the gene *Rp3* in maize that are most easily explained by assuming that the resistance of this gene is dominant against race 901 of *P. sorghi* but recessive against race 903. Lupton and Macer (1962) found that the resistance conditioned by two genes in wheat is dominant against some races of *P. striiformis* but recessive against others. Genetic background, probably through epistasis, can reverse dominance. Working with *P. recondita tritici* Dyck and Samborski (1968) found that in wheat a leaf rust resistance gene from the *Lr2* group was dominant in the background of the wheat variety Red Bobs but recessive in the background of the variety Thatcher.

Epistasis, which includes complementary gene action, is the interaction of one gene with another gene in a different locus. The effect of a gene depends on the presence or absence of some other gene. Epistatic interaction in pathogens is the main theme of Chapter 4, because it is particularly relevant to stabilizing selection.

Recessive resistance has been widely reported in context with complementary gene action. Some caution is needed here. What passes for complementary gene action among recessive genes might often be additive action, with the additive effect of each gene accumulated until enough resistance is reached to be effective. For this reason we choose examples with dominant resistance.

The tomato has independently segregating dominant genes for resistance to *Fusarium oxysporum lycopersici* and to *Verticillium albo-atrum*. Sidhu and Webster (1979) found that, although the gene for resistance to *Fusarium* did not protect against *Verticillium* when the tomato was challenged by *Verticillium* alone, it protected against *Verticillium* when challenged by *Verticillium* and *Fusarium* together. In the F_2 progeny an epistatic ratio emerged because, when both pathogens were used together, the reaction class R-S (resistant to *Fusarium* but susceptible to *Verticillium*) was modified to a phenotypic reaction class R-R (resistant to both pathogens) which was phenotypically indistinguishable from the true reaction class R-R (genetically resistant to both). The gene for resistance to *Fusarium* modified the expression of the recessive allele for susceptibility to *Verticillium* in such a way that it imitated the reaction of its dominant allele for resistance. In an environment that included both pathogens, but not otherwise, there was epistasis.

Another example also involves a single host plant with two pathogens, but with opposite effects. In the tomato the dominant resistance genes to *F. oxysporum lycopersici* and to the rootknot nematode *Meloidogyne incognita* segregate independently. In tomatoes infected with *Meloidogyne* the gene for resistance to *Fusarium* becomes ineffective. In an environment containing both organisms the expression of the gene for resistance to *Fusarium* is modified by the presence or absence of the gene for resistance to *Meloidogyne*, and in a genetic analysis Sidhu and Webster (1974) found segregation ratios appropriate for this epistatic interaction.

A yellow or red pigment in the bulb scales makes onions resistant to the smudge caused by *Colletotrichum circinans*. White bulbs are susceptible. Clarke *et al.* (1944) found that three pairs of genes are involved in the development of pigment: (1) *C-c* is a basic color factor in which the dominant *C* allele is needed before there can be any color at all. The *cc* plants have white bulbs. (2) *R-r* determines color, in the presence of *C*, *R* producing red and *r* yellow scales. (3) *I-i* is an inhibiting factor, with *I* partially dominant over *i*. All *II* plants produce white bulbs. Thus, both *R* and *r* are genes for resistance, their effect being controlled by genes in two other loci.

The importance of genes modifying genes for resistance has recently been stressed for cotton diseases and pests (Bird, 1980). Compatible minor-modifier genes must be present with the resistance genes; this is the key to successful genetic improvement of cotton for disease, insect, and stress resistance. This is a pioneer concept already used to great effect in cotton breeding.

3.3   DISCONTINUOUS VARIATION WITH MANY GENES INVOLVED: PSEUDOMONOGENIC RESISTANCE

Discontinuous variation with many genes involved is a genetic oddity. Variation of resistance in the host plant remains discontinuous even when many resistance genes have been accumulated. This is against the general rule that requires continuity, not discontinuity, when genes determining a trait segregate independently in many loci. This oddity is a feature of gene-for-gene diseases in which resistance genes in the host are matched by corresponding specific genes for virulence in the pathogen.

A single gene pair can determine resistance irrespective of the condition of other gene pairs. There are at least 35 known *Lr* (leaf rust) genes in wheat that condition resistance to *P. recondita tritici* (Browder, 1980). A single effective gene pair *LrLr* can make the phenotype resistant even though there are gene pairs *lrlr* for susceptibility in all other 34 loci. Indeed, because resistance is usually dominant, a single heterozygous pair *Lrlr* can condition resistance; a single dose of resistance can be effective among 69 doses of susceptibility. *P. recondita tritici* is not exceptional; what has been written about it holds commonly for gene-for-gene disease.

The genetic oddity, which keeps the distribution of phenotypes discontinuous despite many independently segregating allelic pairs, is best illustrated by comparing the numerical characteristics of conventional and gene-for-gene dominance systems. Table 3.1 does this. With n allelic pairs there are 3^n possible kinds of genotypes and, with full dominance, 2^n kinds of phenotypes in conventional systems. This is standard genetic theory; see, e.g., Allard (1960, p. 68). But in gene-for-gene systems, irrespective of the number of resistance genes, there is a maximum of only two kinds of phenotype, resistant and susceptible, provided only that the two conditions are met, namely, that resistance is fully dominant and fully effective.

Consider first this maximum and the last column of Table 3.1 and take three allelic pairs (i.e., $n = 3$) as an example for detailed analysis. With R

TABLE 3.1

Oddity of Gene-for-Gene Systems: Numerical Characteristics of Hybrids between Parents Differing in n Allelic Pairs, Full Dominance Being Assumed

		Kinds of phenotypes[a] in F_2	
Number of allelic pairs[b]	Kinds of genotypes possible in F_2	Conventional system	Gene-for-gene system
1	3	2	1[c] or 2[d]
2	9	4	1 or 2
3	27	8	1 or 2
4	81	16	1 or 2
n	3^n	2^n	1 or 2[e]

[a] Environmental variance ignored.

[b] Host genes only.

[c] Resistance genes all made ineffective by matching virulence in the pathogen.

[d] Some or all the resistance genes effective.

[e] The presence of only two kinds of phenotype normally indicates monogenic inheritance with dominance, which is why this sort of resistance is often called monogenic, even when an indefinite number n of allelic pairs is involved.

standing for resistance and *r* for susceptibility, the 27 possible genotypes referred to in Table 3.1 are *r1r1r2r2r3r3* (no resistance); *r1R1r2r2r3r3*, *r1r1r2R2r3r3*, *r1r1r2r2r3R3* (one resistance dose); *R1R1r2r2r3r3*, *r1r1R2R2-r3r3*, *r1r1r2r2R3R3*, *r1R1r2R2r3r3*, *r1R1r2r2r3R3*, *r1r1r2R2r3R3* (two resistance doses); *R1R1r2R2r3r3*, *R1R1r2r2r3R3*, *r1R1R2R2r3r3*, *r1r1R2R2-r3R3*, *r1R1r2R2R3R3*, *r1r1r2R2R3R3*, *r1R1r2r2R3R3* (three resistance doses); *R1R1R2R2r3r3*, *R1R1r2r2R3R3*, *r1r1R2R2R3R3*, *R1R1r2R2r3R3*, *r1R1R2R2r3R3*, *r1R1r2R2R3R3* (four resistance doses); *r1R1R2R2R3R3*, *R1R1r2R2R3R3*, *R1R1R2R2r3R3* (five resistance doses); and *R1R1R2R2-R3R3* (six resistance doses). Of these 27 genotypes only the first conditions a susceptible phenotype. The other 26, given that the resistance genes are unmatched by virulence and remain effective, condition a resistant phenotype.

With four allelic pairs, 1 genotype would condition susceptibility, 80 would condition resistance, and so on. This oddity has led to this sort of resistance being called monogenic. Thus, even though about 40 *Sr* (stem rust) resistance genes are known in wheat and many common wheat cultivars have 5 or more of them, the American Phytopathological Society's "Compendium of Wheat Diseases" can still describe stem rust resistance in wheat as monogenic. This is a realistic appraisal of the oddity, but pseudomonogenic seems to be a more appropriate word.

Consider now the penultimate column of numbers in Table 3.1. In gene-for-gene systems the matching of all resistance genes in the host plant by virulence genes in the pathogen reduces the host to complete susceptibility. However many genes for resistance there were at the start, there is left only one residual phenotype. All phenotypic variance is then environmental. This is the dreaded condition in gene-for-gene disease when epidemics can be expected to occur.

The large number of known resistance genes in gene-for-gene systems is a tribute to the skill and diligence of geneticists. From 20 to 40 genes have been identified for resistance to stem rust and leaf rust of wheat caused by *P. graminis tritici* and *P. recondita tritici*, crown rust of oats caused by *P. coronata avenae*, powdery mildew of barley caused by *Erysiphe graminis hordei*, and rust of flax caused by *Melampsora lini*. Substantial numbers of genes have been identified for most of the fungal diseases known or suspected to be gene-for-gene diseases, and the numbers are growing with further research. Many genes have been identified in cotton for resistance to the bacterial disease caused by *Xanthomonas malvacearum*. Against virus diseases the number of resistance genes on a gene-for-gene basis is necessarily restricted by the small virus genome.

Discontinuity arises from the fact that dominance variance is the major component of genetic variance. On balance, dominant resistance is more

TABLE 3.2

Inheritance of Resistance to Some Diseases of Sorghum[a]

Disease	Pathogen	Inheritance
Anthracnose	*Colletotrichum graminicola*	Dominant
Charcoal rot	*Macrophomina phaseolina*	Recessive
Downy mildew	*Peronosclerospora sorghi*	Dominant
Gray leaf spot	*Cercospora sorghi*	Recessive
Head smut[b]	*Sphacelotheca reiliana*	Dominant, some intermediate or recessive
Milo disease	*Periconia circinata*	Dominant
Rust	*Puccinia purpurea*	Dominant
Bacterial leaf streak	*Xanthomonas holcicola*	Recessive
Bacterial leaf stripe	*Pseudomonas andropogoni*	Recessive
Maize dwarf	Maize dwarf mosaic virus	Dominant

[a] Mostly from Frederiksen and Rosenow (1980).
[b] Possibly a gene-for-gene disease.

is illustrated in Table 3.2. Environmental variance is usually relatively small. Genes of this group have been widely and successfully used in field crops, and with some exceptions resistance has remained stable.

3.5 DISCONTINUOUS VARIATION AND CYTOPLASMIC INHERITANCE

With the coming of hybrid seed and the concomitant difficulty of mechanical and chemical emasculation a search began for extrachromosomal genes conditioning male sterility, i.e., for cytoplasmic genes. In maize the most popular cytoplasmic male sterility was discovered at the Texas Agricultural Experiment Station, and Tms (Texas male sterile) cytoplasm became widely used in making hybrid maize.

The first indication that maize based on Tms cytoplasm was vulnerable to disease came from the Philippines. Mercado and Lantican (1961) reported that Tms maize was very susceptible to *Helminthosporium maydis* (*Cochliobolus heterostrophus*), but at that time there was no warning of danger in the United States. In 1969 Scheifele *et al.* (1969) reported that yellow leaf blight caused by *Phyllosticta zeae* had increased in the United States where Tms lines of maize were used. By 1969, up to 90% of the field maize grown in the United States was based on Tms cytoplasm. In 1970 an epidemic of *H. maydis* swept through the Tms-based fields of the eastern United States. It started in Florida and moved to Georgia and west around the gulf coast of Texas. Northwards it went to Wisconsin and southern Minnesota, and along the coastal plains of Georgia and the Carolinas to New England. Losses of up to 50% were reported. The epidemic was produced by race T of *H. maydis*, and only plants based on Tms or related types of cytoplasm were severely attacked by this race. Normal male-fertile cytoplasm has considerable resistance.

The great 1970 epidemic of *H. maydis* in maize is often quoted as a cautionary example of the danger of genetic uniformity in crops. There is some questionable logic here. Cytoplasmic uniformity is the rule in crops. New cultivars normally start from a single seed and are cytoplasmically uniform. They are as often as not bred from previous cultivars used as maternal parents, thereby perpetuating cytoplasmic uniformity. Millions of acres of field crops have uniform cytoplasm. It was not uniformity but diversification into Tms cytoplasm that caused the 1970 epidemic of *H. maydis*, and so far as plant pathology is concerned there is nothing to suggest that we would be safer if the cytoplasm of our crops like wheat, rice, and sugarcane was more diversified.

3.6 CONTINUOUS VARIATION

Continuously variable resistance is widely used in plant breeding. It is often wrongly called polygenic.

The keys to continuous distribution of resistance are additive variance and nonheritable (environmental) variance. Distributions are continuous because most of the heritable variance is additive or most of the phenotypic variance is nonheritable. The likelihood that it is continuous because the inheritance is polygenic is, on present experimental evidence, small, although there is scope for argument about words.

Suppose that there are three pairs of resistance/susceptibility alleles and that genetic variance is entirely additive. We write R for resistance and S for susceptibility. The 27 kinds of genotype to be found in the F_2 of hybrids between susceptible and resistant parents are *S1S1S2S2S3S3* (no resistance); *S1R1S2S2S3S3, S1S1S2R2S3S3, S1S1S2S2S3R3* (one resistance dose); *R1R1S2S2S3S3, S1S1R2R2S3S3, S1S1S2S2R3R3, S1R1S2R2S3S3, S1R1-S2S2S3R3, S1S1S2R2S3R3* (two resistance doses); *R1R1S2R2S3S3, R1R1-S2S2S3R3, S1R1R2R2S3S3, S1S1R2R2S3R3, S1R1S2S2R3R3, S1S1S2R2-R3R3, S1R1S2R2S3R3* (three resistance doses); *R1R1R2R2S3S3, R1R1-S2S2R3R3,S1S1R2R2R3R3, R1R1S2R2S3R3, S1R1R2R2S3R3, S1R1S2-R2R3R3* (four resistance doses); *S1R1R2R2R3R3, R1R1S2R2R3R3, R1R1-R2R2S3R3* (five resistance doses); and *R1R1R2R2R3R3* (six resistance doses). If the genes have unequal effects, the 27 kinds of genotype condition 27 kinds of phenotype. The 27 kinds of genotype listed here are the same as the 27 kinds listed in Section 3.3, with a change of notation from R and r to R and S to emphasize the change from dominance to additive variance. A change from dominance to additive variance has increased the number of phenotypes from 2 to 27. Plant pathologists are used to scales of rating disease or resistance, e.g., 1 = healthy, 10 = severely diseased; or 1 = resistant, 10 = very susceptible. Few scales have more than 10 classes, many have less. That is, as a matter of experience there is a limit to useful, observable discontinuity, the practical limit being about 10 kinds. As many as 27 kinds represent continuity for all practical purposes. The change from dominance to additive variance is in itself, without help from nonheritable variance, enough to change the variation of resistance from strongly discontinuous to practically continuous.

To consider autotetraploids like the potato instead of diploids, even digenic inheritance is enough, without the help of nonheritable variance, to condition continuous variation. With digenic inheritance there are 25 kinds of genotypes, which, if variance is additive and the genes have unequal effects, condition 25 kinds of phenotypes.

Nonheritable variance makes for continuous distribution even if the genetic variance conditions discontinuity. Continuously variable phenotypes can accompany even monogenic resistance if nonheritable variance exceeds heritable variance. Resistance to cucumber mosaic was thought to be "polygenic" until tests were made under more strictly controlled conditions, when it was found to be monogenic (Walker, 1966). The ratio of nonheritable to heritable variance needed to make monogenic resistance continuously variable varies, the difference in resistance between the parents being a factor. See, e.g., Allard (1960, p. 48). When resistance is additive, one can usually expect continuously variable phenotypes when nonheritable exceeds heritable variance, irrespective of the number of resistance genes involved.

Nonheritable variance in this group of diseases is usually great, and specially sensitive to varying amounts of inoculum and varying conditions of humidity and temperature that affect inoculation.

Continuously variable resistance is commonly ascribed to polygenic inheritance. The experimental evidence is against this, and no more resistance genes of substantial effect seem to be involved here than those that condition discontinuous resistance discussed in Section 3.3. In other words, as a cause of continuous variation of resistance, polygenic inheritance is probably a poor third behind additive variance and environmental variance. The crucial question is, how many genes determine the bulk of the variation?

Reciprocal chromosomal translocation has been used to count resistance genes. Consider the work of Jenkins and Robert (1961) with leaf blight in maize caused by *Helminthosporium turcicum* (= *Exserohilum turcicum*). The form of resistance they studied is continuously variable (unlike the *Ht* form). It is the form commonly called polygenic. The inbred line Mo21A has high resistance. Its resistance was transferred and used to develop the resistant lines CI. 28A, CI. 42A, and CI. 64 from susceptible lines, by first crossing the susceptible lines with Mo21A and then backcrossing to the susceptible lines for two or three generations. Jenkins and Robert determined the number and position of the resistance genes in Mo21A and how many of them were transferred and needed for resistance in CI. 28A, CI. 42A, and CI. 64. Another resistant line CI. 9A of independent origin was also included in their tests.

In Mo21A genes for resistance are located in 2L (the long arm of chromosome 2), 3L, 4S (the short arm of chromosome 4), 4L, 5L, 6S, 6L, and 7S. Of these, the genes located in 5L and 7S are the most important genes for resistance. They yielded some of the largest differences between mean blight scores. The gene located in 3L is also important. These genes, located in 5L, 7S, and 3L, were all recovered in the derived resistant lines, CI. 28A, CI. 42A, and CI. 64. They were also found in CI. 9A. No other resistance gene was identified in CI. 28A; from the evidence, this line is simply trigenic for resistance. The gene in 2L was recovered in CI. 42A but not CI. 64, and the genes in 4S and 4L were recovered in CI. 64 but not CI. 42A. The gene in 6L was not recovered from any of the derived lines, and evidence about the gene in 6S is lacking because of an experimental mishap. (It was not identified in CI. 9A.) The genes in the different locations evidently have effects of different sizes, with three genes (or genes in three locations) outweighing the others. When genes differ in the size of their effects, the total number of genes, which is the issue in polygenicity, becomes of subsidiary interest.

These results of Jenkins and Robert (1961) are supported by the work of Hughes and Hooker (1971) who concluded that resistance in maize to *H. turcicum* is inherited simply and conditioned by few genes; they suggested from three to six. Additive gene effects were of major importance in all populations in both years of study.

To cite some other literature about gene numbers conditioning continuously variable resistance to disease in maize, Grogan and Rosenkranz (1968) found that resistance in maize to maize stunt virus seemed to be entirely additive. They could not accurately determine the number of genes involved; the indications were that the inheritance was simple, but with more than 1 gene pair conditioning resistance. Kim and Brewbaker (1977) worked with the continuously variable form of resistance in maize to rust cuased by *Puccinia sorghi*. They used two inbreds highly resistant to rust and crossed them with susceptible inbreds. They estimated that one of the resistant inbreds Oh545 had 2 gene pairs conditioning resistance. Estimates for the other resistant inbred Cm105 by three different methods averaged

1.3 gene pairs, which presumably indicates 1 pair of large effect and others of less effect. Carson and Hooker (1982) studied anthracnose stalk rot of maize caused by *Colletotrichum graminicola*. They concluded that resistance in the highly resistant inbred A556 was mostly additive and may be controlled by as few as 5 genes.

To get these estimates of numbers in perspective one might draw comparisons with discontinuously variable resistance conditioned by pseudo-monogenes and discussed in Section 3.3. The wheat cultivars Selkirk and Canthatch each have at least five *Sr* genes for resistance to *P. graminis tritici*. There is nothing to suggest that this is unusually many for wheat cultivars. With roughly the same number of genes involved, resistance to the maize diseases discussed in this section is commonly called polygenic, and resistance to the wheat disease and others like it monogenic. As prefixes, "poly" and "mono" have come to have little to do with gene numbers in plant pathology.

3.7   BACKGROUND TO THE POLYGENE STORY

The rediscovery of Mendel's laws at the beginning of the century started one of the great controversies in the history of biological science. Darwinism and Mendelism were thought to be conflicting opposites. Darwinism was upheld by the biometric school, Mendelism by those who saw mutations as the starting point of new species. In Darwinism, evolution was seen in terms of continuous variation reflecting very small differences that natural selection could in the course of time accumulate to start new species. Mendelian mutants they considered to be irrelevancies which natural selection would eliminate. In Mendelism, discontinuity in the form of mutants that could be seen to obey Mendel's laws of inheritance was the stuff new species were made of. The debate was largely settled by the work of population geneticists, notably Fisher, Haldane, and Wright. Continuous variation, studied by biometricians, could be explained by alternative alleles at many loci, each by itself having a very small effect on the phenotype. Mather (1943) rounded off the discussion by naming and defining polygenes: polygenic inheritance was by large numbers of genes, each of very small effect. The title of Mather's paper, "Polygenic inheritance and natural selection," reveals the context in which polygenes were defined.

Population geneticists and plant breeders think differently of time. Population geneticists pondering the working of natural selection can comfortably contemplate the origin of species a million or more years in the making. Genes of very small effect, polygenes, are sufficient for this. Plant breeders are hastier; they want results within a few years. Only genes of large effect, whether evident through discontinuous variation or concealed in continuous variation, are sufficient for this. Polygenes in Mather's sense are not the material of new cultivars.

3.8   THE ERROR OF EXPECTING SAFETY IN NUMBERS: ADDITIVE VARIANCE

Polygenic resistance, it is often argued, is safe resistance that the pathogen cannot match. Safety, it is reasoned, lies in numbers. The more resistance genes the pathogen has to match, the less likely it is to match them. Monogenic resistance, the reasoning runs, is unstable because the pathogen can, by a single mutation, add virulence and match the resistance. Polygenic resistance requires a multitude of matching virulences that the pathogen cannot accomplish, and the resistance is stable. Such reasoning must be rejected entirely. Indeed it has become hypothetical, because it is now reasonably certain that resistance used by plant breeders is seldom, if ever, polygenic. In any case the idea of safety in numbers is simply against the facts. On the one hand, much monogenic or oligogenic resistance has remained stable over the years. On the other, multiple virulence that can match many resistance genes is common. Most isolates of *P. graminis tritici* in Canada are virulent for at least 10 *Sr* genes: *Sr5, 7a, 7b, 8, 9d, 9e, 10, 11, 14*, and *Tt1*. What is more, much of this virulence occurs even in the absence of the corresponding resistance genes; several of the *Sr* genes just named, like *Sr10*, are not known to occur in the wheat fields of North America. Virulence by association is discussed in Chapter 4.

Nevertheless, there is a matter for investigation. It concerns additive variance, not polygenic inheritance. The one has been mistaken for the other. There is a substantial amount of literature to suggest that continuously variable (i.e., quantitative) resistance inherited additively often has a record of stability in the face of pathogenic variability. Many examples come from maize diseases. Much of the resistance used by maize breeders is continuously variable, much is additively inherited, and much has stood the test of time. The present level of resistance of North American maize hybrids effectively controls disease. As Hooker (1979) has pointed out, the 1975 season was wet and should have favored many leaf diseases. In fact, few disease problems were encountered, and grain yields were high. This resistance has been accumulated over the years, without lapses and setbacks, and its accumulation has been primarily additive. Consider two examples.

Resistance to rust in maize caused by *P. sorghi* is largely additive (Kim and Brewbaker, 1977). Rust is potentially destructive under moist conditions, as can be seen in susceptible hybrids. In them, losses in grain yield of 6.3–23.5% have been reported (Hooker, 1962; Russell, 1965). However, resistance has kept losses low. An adequate level of resistance exists within the germ plasm pool of the American corn belt (Hooker, 1979). When inbred lines were first developed in the United States, maize breeders selected heavily against rust. The resistant lines they selected provided in due course the germ plasm base for the inbred lines now in use, and the resistance is still there (Hooker, 1979). There is enough resistance to keep grain losses from rust small. This resistance is inherited additively, and has been maintained over the years.

Resistance has remained stable despite the cultivation of maize over immense and continuous areas, and despite the variability of *P. sorghi* aided by a sexual stage on *Oxalis* spp. which are often heavily infected in Mexico (Borlaug, 1946). Uredospores of *P. sorghi* travel far, and there would have been ample scope for the spread of new pathogenic races had they developed. Yet there have been no reports of maize hybrids having had to be changed because of new races of *P. sorghi*.

Resistance to leaf blight in maize caused by *H. turcicum* is inherited additively (Hughes and Hooker, 1971). When maize inbreds were first being developed, insufficient attention was given to blight resistance and many of the inbred lines were highly susceptible. In hybrids many loci became homozygous for susceptibility where previously open-pollinated crops were heterozygous and heterogeneous. In 1942, at the time when hybrid maize was replacing open-pollinated maize, a destructive epidemic developed. In 1942, Elliott and Jenkins (1946) tested 38 commercial inbred lines of corn belt maturity at Beltsville, exposing them to natural infection. By August 12, without artificial inoculation, 15 of these lines had abundant lesions on the lower leaves. Only 12 of the lines had no more than slight infection by August 12. The general high level of susceptibility of the new inbred lines of corn belt maturity explained the 1942 epidemic; the change from open-pollinated to hybrid maize had not been adequately prepared for. New lines were selected for resistance during the following years. This resistance was adequate. It has persisted, even though the fungus is variable, with different forms pathogenic to maize, Sudan grass, sorghum, and Johnson grass and with large differences in aggressiveness between cultures. Pathogenic variability has not led to a loss of leaf blight resistance in agriculture (Hooker, 1979), and there is no evidence to suggest that the resistance accumulated by breeders since 1942 can be negated by pathogenic variation.

3.9   EXPERIMENTAL DIFFICULTIES IN PARTITIONING VARIANCE

Variance is additive if a hybrid is midway between its homozygous parents. Departures from the midpoint indicate nonadditive effects. A sophisticated biometry exists about all this. Our concern here is with the experimental evidence. In plant disease, what is a midpoint?

Most experiments to determine additive and nonadditive effects have used scales of rating, e.g., 1 = no disease, 10 = severe disease. Can we be sure that, say, a rating of 5 is midway between ratings of 4 and 6, or 3 and 7? There is nothing to show that the scales were so constructed or could have been so constructed. Indeed, there is an inherent difficulty. The best scales are logarithmic, according to the Weber–Fechner law (Horsfall and Cowling, 1978); the needed midpoints are arithmetic.

Disease has also been measured as the percentage of plants infected. If 10% of the plants of the resistant parent and 90% of the plants of the susceptible parent were infected, would 50% be the midpoint? It is not. If infections were randomly distributed, there would be an average of 105 infections per 1000 plants of the resistant variety, and 2303 infections per 1000 plants of the susceptible parent. The midpoint would be 1204 infections per 1000 plants, equivalent (if infections were randomly distributed) to 70% of the plants being diseased. That is, 70% is the midpoint between 10 and 90%, infections being random. Nonrandomness adds its own complication.

Several authors in partitioning variance have equated susceptibility with the rate of spread of the lesion from the point of inoculation. If in the resistant parent the lesion spread 1 mm a day and in the susceptible parent 3 mm a day, would 2 mm a day be the midpoint? Why should it be? Should one rather take areas, and find the midpoint between, say, 1 and 9 mm^2 per day? With rots, should one rather take volumes and find the midpoint between, say, 1 and 27 mm^3 per day? One can ring the changes and count, say, the number of internodes of stalk traversed by the pathogen in a given time, but the difficulty of interpretation remains.

What all this comes to is that the literature often underestimates additive variance and overestimates nonadditive variance. Indeed, several reports of dominance and epistasis must be treated with skepticism. It also means that good biometry cannot compensate for incorrect measurements or interpretation of measurements.

4

The Pathogen: Epistasis and Virulence

4.1 INTRODUCTION

This chapter changes the emphasis from the host to the pathogen. It deals primarily with epistatic interaction and the structure of virulence. This in turn leads to an examination of the genetic basis of stabilizing selection in populations of the pathogen and to a clearer understanding of physiological races.

Epistasis implies interactions between genes at different loci, whereby the effect of one gene is changed by the presence or absence of other genes. Epistasis is interallelic interaction, i.e., the interaction of nonalleles, as distinct from intraallelic interaction on which dominance variance discussed in Chapter 3 is based.

In populations of *Puccinia graminis tritici* in North America virulence for the gene *Sr6* is common, as is virulence for the gene *Sr9e*, but virulence for *Sr6* and *Sr9e* in combination is rare. A wheat cultivar with *Sr6* alone or *Sr9e* alone would be susceptible to stem rust, but (as things are at present in North America) a cultivar with both these genes would be resistant. Consider the pathogen. On the assumption that there is a gene-for-gene relation between wheat and *P. graminis tritici*, the effect of the gene for virulence for *Sr6* is evidently changed by the presence or absence of the gene for virulence for *Sr9e*. This is epistasis in the pathogen. Consider the host. Here too there is epistasis, by reflection from the pathogen. The effect of the gene *Sr6* in wheat is changed by the presence or absence of the gene *Sr9e*, via the behavior of the pathogen.

With epistasis goes stabilizing selection against the accumulation of virulence for the two *Sr* genes. Stabilizing selection is the opposite of directional selection or adaptation. Without opposite selection pressures, *P. graminis* would adapt itself to a wheat variety with both *Sr6* and *Sr9e* by accumulating associated virulence for these two genes. (No new mutations are needed; associated virulence is already known in North America, e.g., in Canadian formula C50.) This tendency of these virulences to dissociate, because of epistasis, would act to stabilize the resistance of wheat with the two genes, the action being proportional to the epistatic interaction.

There is also an opposite epistatic interaction promoting the association of virulence. This can have, and as a matter of history probably has had, a destabilizing effect on resistance. In any case, associated virulence is just as important as dissociated virulence in determining the sort of population a pathogen will have.

Puccinia graminis tritici is the vehicle for discussion in most of this chapter because of the relatively abundant data that are available.

4.2   THE ABC–XYZ CLASSIFICATION AND DIALLEL GENE PAIRING

The *Sr* genes in wheat for resistance to stem rust can be divided into groups (Vanderplank, 1982). There is an ABC group which on North American evidence comprises *Sr6*, *9a*, *9b*, *15*, and *17*; an XYZ group comprising *Sr7b*, *9e*, *10*, *11*, *Tt1*, and *Tmp*; and an N (neutral, nondescript, nonaligned) group which is either neutral or simply unclassified for lack of evidence. Virulence for any *Sr* gene in the ABC group tends to dissociate from virulence for any *Sr* gene in the XYZ group, i.e., the actual frequency of combined virulence for a pair of *Sr* genes, one from each group, is less than the theoretical frequency calculated to occur if the combinations were random. Correspondingly, virulence for any *Sr* gene in either group tends to associate with virulence for any other *Sr* gene of the same group, i.e., the actual frequency of associated virulence for a pair of *Sr* genes both from the ABC group or both from the XYZ group is greater than the theoretical virulence calculated for random combination.

The method of diallel gene pairing introduced by Roelfs and Groth (1980) can be adapted to probe the ABC–XYZ system. Table 4.1 uses it, rearranging Roelfs and Groth's data on an ABC–XYZ basis. The data are for 1975.

TABLE 4.1

Percentage on the Diagonal of Isolates of *Puccinia graminis tritici* Virulent for Selected
Sr Genes, and above and below the Diagonal the Theoretical and Actual Percentages of
Virulence for Diallel Pairs of These Genes[a]

Theoretical / Actual	Percentages of isolates virulent for indicated Sr gene(s)										
	ABC group					XYZ group					
	6	9a	9b	15	17	7b	9e	10	11	Tt1	Tmp
6	**11**	2	1	2	2	9	8	10	9	9	8
9a	2 =	**14**	2	3	3	11	11	12	11	11	11
9b	3>	4>	**12**	3	2	10	9	10	10	10	9
15	11>	14>	13>	**23**	5	19	18	20	19	19	18
17	9>	10>	10>	20>	**20**	16	15	18	16	16	15
7b	2<	4<	4<	6<	2<	**82**	63	72	67	66	63
9e	0<	*<	*<	*<	*<	77>	**77**	68	63	63	59
10	9<	2<	10 =	11<	11<	80>	77>	**88**	72	72	68
11	10>	2<	11>	11<	8<	74>	72>	80>	**82**	66	63
Tt-1	4<	4<	4<	4<	1<	81>	77>	78>	74>	**81**	62
Tmp	0<	0<	0<	*<	*<	77>	77>	77>	72>	77>	**77**

[a] Data of Roelfs and Groth (1980) for 2377 isolates from wheat and barley in the United States in 1975, rearranged in ABC–XYZ groups. The >, <, and = signs show whether the actual percentage is greater than, less than, or equal to the theoretical. * indicates less than 0.6% virulence.

Diagonal entries show the percentage frequency of virulence for the *Sr* genes individually. Thus, 11% of the isolates were virulent for *Sr6* and 77% for *Sr9e*. Entries above the diagonal show the expected percentage frequency of virulence for each pair of genes if virulence is distributed randomly. Thus, $11 \times 77/100 = 8\%$ of the isolates would, if randomly distributed, be virulent for both *Sr6* and *Sr9e*. Entries below the diagonal show the actual frequency of virulence for *Sr* gene pairs. Thus, 0% of the isolates were actually virulent for both *Sr6* and *Sr9e*. Because this is less than the theoretical figure 8%, the entry is marked with a < sign.

Consider the 55 entries of actual virulence frequencies, i.e., those below the diagonal. The block of 30 entries in the bottom left-hand corner of the table is for paired virulence for *Sr* genes, one from the ABC and one from the XYZ group. According to the ABC–XYZ classification these entries should have < signs. The remaining 25 entries below the diagonal for paired virulence for *Sr* genes both from the ABC or both from the XYZ groups should have > signs. This is essentially what is found. There are two reversals, both trivial in size and both related to *Sr11*. That is, this gene's membership of the XYZ group was loose in 1975.

TABLE 4.2

Percentage on the Diagonal of Isolates of *Puccinia graminis tritici* Virulent for Selected *Sr* Genes, and above and below the Diagonal the Theoretical and Actual Percentage for Diallel Pairs for these Genes[a]

Theoretical ⟍ Actual	Percentage of isolates virulent for indicated *Sr* gene(s)									
	ABC group					XYZ group				
	6	9a	9b	15	17	9e	10	11	Tt1	(9dH)[b]
6	**14**	1	1	3	1	12	14	13	12	13
9a	4>	**7**	*	1	1	6	7	6	6	6
9b	6>	5>	**6**	1	1	5	6	5	5	6
15	14>	7>	7>	**18**	2	15	18	16	16	17
17	5>	3>	5>	6>	**10**	8	10	9	9	9
9e	1<	2<	2<	2<	2<	**84**	81	76	74	79
10	12<	4<	5<	15<	6<	84>	**97**	87	85	91
11	11<	3<	5=	12<	5<	79>	89>	**90**	79	85
Tt1	3<	4<	4<	5<	2<	85>	85=	80>	**88**	83
(9dH)[b]	1<	2<	2<	2<	2<	83>	93>	87>	95>	**94**

[a] Data of Green (1976a) for 332 isolates from wheat, barley, and rye in Canada in 1975, arranged in ABC–XYZ groups. Some of the data, especially those covered by formula C25, were incomplete, and adjustments had to be made. But the adjustments have no noteworthy effect on the entries in the table. The >, <, and = signs show whether the actual percentage is greater than, less than, or equal to the theoretical. * indicates less than 0.5% virulence.

[b] Recorded as *Sr9d* in the survey, but two genes *Sr9d* and *SrH* are involved (Green, 1981). Gene *H* is probably the XYZ gene.

Table 4.2 gives corresponding data from Canada, also for 1975. Compared with Table 4.1 there are two omissions. Gene *SrTmp* is not used in the Canadian surveys, and *Sr7b* entered the surveys only after 1975 when, as in the United States, it was seen to belong to the XYZ group (Vanderplank, 1982). For the rest, agreement with the United States survey and the ABC–XYZ classification is good.

Gene *Sr17* appears as a member of the ABC group in both the United States and Canadian surveys in 1975. In other years virulence for it behaves highly erratically, and the gene's membership is in doubt. To pursue this detail would, however, be out of place in this book.

4.3 VIRULENCE DISSOCIATION

Virulences behave differently in epistatic dissociation. Of the XYZ group of *Sr* genes, virulence for *Sr9e* and *SrTmp* is (on present evidence) the most

TABLE 4.3

Percentage of Isolates, Theoretical and Actual, of *Puccinia graminis tritici* Virulent
in Canada for Wheat Stem Rust Resistance Gene *Sr9e* in Combination
with *Sr6, Sr9a, Sr9b,* and *Sr15* in the Years 1974 to 1979[a]

Resistance genes	Year					
	1974	1975	1976	1977	1978	1979
Sr9e and *Sr6* theoretical[b]	9	12	5	17	13	10
Sr9e and *Sr6* actual[c]	0	*[d]	0	0	0	0
Sr9e and *Sr9a* theoretical	7	6	5	4	3	18
Sr9e and *Sr9a* actual	*	2	0	*	0	0
Sr9e and *Sr9b* theoretical	12	5	6	18	12	12
Sr9e and *Sr9b* actual	*	2	0	*	0	0
Sr9e and *Sr15* theoretical	15	15	7	20	22	25
Sr9e and *Sr15* actual	2	2	*	*	3	1

[a] Data of Green (1975, 1976a,b, 1978, 1979, 1980). The number of isolates was 429, 332, 602, 653, 304, and 288 for the 6 years, respectively.
[b] Theoretical percentage if the virulence were randomly distributed.
[c] Actual percentage in the surveys.
[d] * = less than 1%.

TABLE 4.4

Percentage of Isolates, Theoretical and Actual, of *Puccinia graminis tritici* Virulent in Canada
for Wheat Stem Rust Resistance Genes *Sr6, Sr9a, Sr10, Sr11,* and *Sr15* in Appropriate Pairs
from 1970 to 1978[a]

Sr genes	Year								
	1970	1971	1972	1973	1974	1975	1976	1977	1978
6 and 10 theoretical[b]	8	14	13	9	7	14	5	22	16
6 and 10 actual[c]	*[d]	0	1	0	7	12	1	22	17
6 and 11 theoretical	8	13	12	9	10	13	4	19	15
6 and 11 actual	*	*	2	*	8	11	1	22	17
9a and 10 theoretical	10	20	6	20	8	7	5	5	3
9a and 10 actual	1	0	0	0	2	4	*	3	1
9a and 11 theoretical	10	14	13	10	6	6	4	5	15
9a and 11 actual	3	*	1	2	2	3	*	1	*
15 and 10 theoretical	17	21	15	9	16	18	6	27	27
15 and 10 actual	9	7	4	*	11	15	2	25	21
15 and 11 theoretical	17	21	15	10	15	16	6	23	25
15 and 11 actual	6	5	4	3	10	12	1	24	20

[a] Data of Green (1971, 1972a,b, 1974, 1975, 1976a,b, 1978, 1979).
[b] Theoretical percentage if virulence were randomly distributed.
[c] Actual percentage found in the surveys.
[d] * = less than 1%.

clearly involved in epistatic interaction. Table 4.3 reveals how year after year in Canada virulence for gene *Sr9e* strongly dissociates from virulence for genes of the ABC group, *Sr6*, *Sr9a*, *Sr9b*, and *Sr15*. United States findings are similar for gene *Sr9e* and equally include *SrTmp*.

Table 4.4 deals with virulence for two members of the XYZ group, genes *Sr10* and *Sr11*, that are less consistent members of the group than *Sr9e* and *SrTmp*. Virulence for them also dissociates from virulence for genes *Sr6*, *Sr9a*, and *Sr15* of the ABC group, but the dissociation is more erratic from year to year.

The year 1975 chosen for Tables 4.1 and 4.2 understates rather than overstates the evidence for dissociation in that dissociations revealed in Tables 4.3 and 4.4 were in general less marked in 1975 than in many other years. The tables do not exaggerate.

4.4 EPISTASIS–ENVIRONMENT INTERACTION

Environment strongly affects epistatic interaction. Results vary largely from year to year. Table 4.4 illustrates this for Canada. In 1970 through 1973 virulence for gene *Sr6* and *Sr10* was strongly dissociated; in 1974, 1977, and 1978 there was no evidence for dissociation. Similar comments apply to other gene pairs, especially *Sr10* and *15* or *Sr11* and *15*.

The relevant environmental effects are those preceding the yearly surveys. Temperature variation is in theory important, higher temperatures promoting epistatic interaction, but the literature is uninformative. Katsuya and Green's (1967) experiments are interesting and in the expected direction because higher temperatures reduced the fitness of race 15B (Can) of *P. graminis tritici* relative to race 56. But the experiments concerned artifacts (see Section 4.5) and possibly do not tell the full story. Environmental changes within seasons, from winter to summer, are far greater than those between seasons. Roane *et al.* (1960), working in the field with *P. graminis tritici* in Virginia, found that the ratio of race 15B to race 15 dropped as the season advanced from spring to summer. This finding accords with expectation, for whereas the ABC–XYZ classification is unclear in race 56 it is sharply clear in race 15B. The findings also highlights the point that erratic survey results must inevitably follow erratic dates of survey.

The method of diallel gene pairing is a great advance on the system of representing a pathogen population as a mixture of fixed physiological races. Virulence for diallel gene pairs changing in frequency with change in time would allow changes in the population to be seen in terms of gene flow, thereby bringing plant pathology into line with modern genetics. To be meaningful, surveys would need to include data on time (either in terms of

growth stages of the host or as calendar dates), temperature at or preceding
the date of sampling, and other environmental factors considered to be
relevant. A computer could handle the analysis.

4.5   DANGER IN ARTIFACTS

Epistasis, a commonplace topic in genetics and so evidently concerned in
virulence dissociation, would be expected to have a large share of the literature of virulence structure. In fact, epistasis is seldom mentioned in this
context. The fault may lie with plant pathologists' penchant for artifacts.
Problems that belong to the field are brought into the laboratory or greenhouse, and the results obtained in the laboratory or greenhouse are those
that are believed, even when they are flatly at variance with results from
outside.

Flor (1953, 1956) long ago warned about artifacts and documented
contradictions between observations in the field and greenhouse. In the field
unnecessary virulence reduced the fitness of the pathogen. Flor analyzed
field surveys of the flax rust fungus *Melampsora lini* from 1931 to 1951. In
1931 the leading flax cultivar in the North Central states of the United States
was Bison. Bison is susceptible to all isolates of *M. lini* that have been collected in North America since surveys started in 1931; it has no resistance
genes relevant to the North American scene. Bison was largely replaced by
Koto, which became widely grown by 1942, and later by Dakota, which
possessed more resistance. With more resistant flax cultivars came more
virulent races of *M. lini*. Flor noted that throughout the period of this
analysis the most prevalent races of *M. lini* were those possessing the least
number of virulence genes compatible with survival, i.e., compatible with
the ability to attack the popular cultivars. Throughout the period of his
analysis the percentage of isolates of *M. lini* carrying unnecessary genes for
virulence, i.e., genes enabling them to attack cultivars not grown commercially, decreased. As background to this, note that the ability of *M. lini* to
mutate or otherwise vary is not in question. *M. lini* is eu-autoecious; and
because uredospores do not overwinter in the North Central states, sexual
hybridization initiates each year's infection. Selection pressure rather than
restricted variation curbs unnecessary virulence.

In the greenhouse Flor (1953) got very different results. There was no
evidence for selection against unnecessary virulence. Race 22 of *M. lini*,
which possesses at least 20 pairs of recessive virulence genes, sporulated
as abundantly on Bison as did race 1, which possesses none. In the greenhouse there was no evidence for selection against unnecessary virulence;
and results from the artificial condition in the greenhouse would have led

to entirely wrong conclusions about what happens naturally in the field. Flor tentatively ascribed the field results to heterozygote advantage. Virulence is usually recessive, and phenotypic avirulence commonly heterozygous. Flor argued that heterozygosity could help the avirulent phenotype to adapt to the widely ranging temperatures for germination and infection in the field. The argument is in line with genetic opinion that the heterozygote's flexibility in coping with a fluctuating environment is one of the foundations of heterosis. Be this as it may, we can at least concentrate on Flor's direct evidence that results in the field and in the greenhouse differed so much that the greenhouse evidence could be ignored.

To deal more generally with the use of artifacts in studies of relative fitness, one must allow that small environmental differences can have large effects, and that it is normally unknown exactly what the relevant environmental differences are. To draw on Table 4.4 for examples, in 1976 virulence in *P. graminis tritici* for genes *Sr6* and *Sr10* dissociated, as did virulence for genes *Sr6* and *Sr11*, *Sr15* and *Sr10*, and *Sr15* and *Sr11*. In 1977 evidence for corresponding dissociation was absent. Yet one does not know how 1977 differed relevantly from 1976, and attempts to reproduce conditions of 1976 in the laboratory or greenhouse might well end in reproducing conditions of 1977, with contrary results.

Greenhouse and laboratory experiments are dangerous in matters relating to the comparative fitness of phenotypes to survive in populations of pathogens in a fluctuating environment. Results from them will be quoted only when there is corroborative evidence for their validity.

4.6 VIRULENCE ASSOCIATION

Tables 4.1 and 4.2, through the > signs, show how virulence for genes all in the ABC group tends to associate, as does virulence for genes all in the XYZ group. Some of the associations are tight, e.g., in the XYZ group between virulence for *SrTmp* and virulence for *Sr7b*, *Sr9e*, *Sr10*, and *SrTt1*.

Usually, as a generalization virulence associations are adaptations to resistance genes in the host. If the host plant has resistance genes *R1* and *R2* the pathogen is likely to have associated virulence for these genes. Exceptionally, little or no adaptation of this sort is found. It is not found in *P. graminis tritici* in the wheat fields of the United States and Canada. The stem rust surveys in North America are therefore specially informative; they reveal the internal structure of stem rust virulence with few complications from the pathogen's adaptation to the host.

Table 4.5 shows the distribution in 1975 of wheat cultivars that are field susceptible to *P. graminis tritici* in the United States east of the Rocky

TABLE 4.5

Wheat Cultivars Susceptible in 1975 to *Puccinia graminis tritici* in the United States
East of the Rocky Mountains, and Their *Sr* Genes[a]

Cultivar	Relevant *Sr* gene(s)[b]	Acreage (%)[c]	Region
Sturdy	None	0.9	Central Texas
Knox type	None	0.2	Central Texas
Monon	None	1.1	Southern and Ohio Valley
Yorkstar type	None	1.0	Southern and Ohio Valley
Triumph type	*Tmp*	6.8	Southern Great Plains
Tascosa-Sturdy-Wichita type	None	8.5	Southern Great Plains
Yorkstar type	None	0.3	Northeastern states
Blueboy	None	0.2	Northeastern states
Parker	None	0.2	Upper and mid-Mississippi
Winalta[d]	17, *Tmp*	1.8	Northwestern Great Plains
Cheyenne	None	1.6	Northwestern Great Plains
Warrior	None	0.3	Northwestern Great Plains
Winoka	*Tmp*	0.3	Northwestern Great Plains
Total		23.2	

[a] Extracted from the data of Roelfs and Groth (1980). Susceptible cultivars are those recorded as susceptible in the field. Field resistant cultivars, omitted from the table, may owe their resistance to genes other than those mentioned in Tables 4.1 and 4.2. E.g., the resistant cultivar Eagle owes its resistance to gene *Sr26*.

[b] *Sr* genes for seedling resistance included in Tables 4.1 and 4.2.

[c] Acreage of susceptible cultivars in 1975 as a percentage of the total wheat acreage east of the Rocky Mountains.

[d] The contradiction of Table 4.1 is more apparent than real, because only three of the isolates of *P. graminis tritici* were from the northwestern Great Plains.

Mountains. These susceptible cultivars covered 23% of the total wheat acreage, the remaining 77% of the acreage being covered by resistant cultivars that did not affect the rust population. As Roelfs and Groth (1980) point out, the wheat cultivars do not directly explain the frequency of virulence in the rust population. For example, the most frequent virulence recorded in Table 4.1 is that for the gene *Sr10*, yet, so far as is known, gene *Sr10* is not in any United States wheat cultivar, resistant or susceptible. There is no direct clue in Table 4.5 to the pairing of virulences in Table 4.1.

There is an indirect clue. The gene *SrTmp* occurred in almost 2/5 of the susceptible wheat acreage, being common in the southern Great Plains. No other relevant *Sr* gene was common in the susceptible cultivars. By the association of virulences apparent in Table 4.1, virulence for gene *SrTmp* could have introduced virulence for other genes in the XYZ group. This might account for the predominance of virulence for the XYZ group in

P. graminis tritici from the southern Great Plains northward. To put the suggestion differently, virulence for the gene *SrTmp* might be the only important direct adaptation of pathogen to host recorded in Table 4.1, epistatic interaction inducing indirectly the associated virulences for other members of the XYZ group.

Before 1950 the dominant race of *P. graminis tritici* in the northern Great Plains was race 56. It was avirulent for most of the *Sr* genes of the XYZ group. It was succeeded, accompanied by great epidemics of stem rust in the early 1950s by race 15B, which is virulent for genes of the XYZ group. Some historical research is needed to determine whether the gene *SrTmp* can be blamed for the change, i.e., whether there was destabilizing selection initiated by virulence for the gene *SrTmp*.

Just as virulence dissociation is something useful for the plant breeder to aim at, virulence association that accumulates virulence is harmful. It might well be a waste of effort, if not perhaps positively harmful, for a plant breeder to introduce into a new cultivar a resistance gene when another of the same group, ABC or XYZ, is already present.

4.7 DEFINITION OF A PHYSIOLOGICAL RACE

A physiological race is a nonrandom assemblage of virulences and avirulences, determined externally by adaptation to the host plants and internally by interallelic and intraallelic interactions, all subject to environmental control.

The assemblage must be nonrandom; otherwise there would be no recognizable races. For example, Canadian surveys of *P. graminis tritici* have determined virulence or avirulence for 16 *Sr* genes (*Sr5, 6, 7a, 7b, 8, 9a, 9b, 9d, 9e, 10, 11, 13, 15, 17, Tt1,* and *Tt2*). With randomness there would be $2^{16} = 65,536$ races, with equal expected frequencies. Actually, only 20–30 races of very unequal frequency are usually recorded in any one season, even though hundreds of isolates are analyzed. Races have a nonrandom structure. The comments that follow concern the sources of this structure.

Adaptation to the host, if substantial, leads to the period of bust in a boom-and-bust cycle. Adaptation is the most frequently discussed feature of physiological races and likely to swamp other determinants of structure. Fortunately, it does not do so in the United States and Canadian wheat stem rust populations, which gives them a special place in the study of internal structure determinants.

Interallelic interaction, or epistasis, has been the topic of most of this chapter. Because of the paucity of other data the discussion has perforce been restricted to pairs of virulences, but the restriction is no part of the definition. Interallelic interaction between a gene in *any* locus in the pathogen and a virulence/avirulence gene in another locus would be epistatic.

Intraallelic interaction enters this chapter only as a passing reference to Flor's (1953) suggestion that heterozygosity might adapt avirulent phenotypes to a fluctuating environment. Dominant avirulence and heterozygous avirulent phenotypes are well known. In the general genetic theory of stabilizing selection (homeostasis), heterozygosity plays a large role (Lerner, 1954, 1958), and intraallelic interaction probably deserves far more attention than the limited available data allow us to give it now.

The relevant environment is not only the physical but also the biotic environment, and not only the present but also the historic past environment, as when summer races reflect the environment of winter, or this year's races reflect last year's.

Race theory is easily converted into gene flow theory. Gene flow in pathogens is nonrandom and determined externally by adaptation to host plants and internally by interallelic and intraallelic interactions, all subject to environmental control.

5

Adaptation of the Pathogen to the Host: Wheat Stem Rust in Australia

5.1 INTRODUCTION

The story of boom-and-bust cycles in breeding for resistance to disease is well known. The plant breeder introduces a resistance gene, generally a pseudomonogene, into a new cultivar; the cultivar becomes popular among farmers because of its resistance; the pathogen adapts itself to the new resistance with a change from avirulence to virulence; and the resistance is "lost." No long list of examples need be given. Indirect adaptations also occur; they are less well known. Australian evidence with wheat stem rust illustrates them. It also complements and counterbalances the North American evidence.

In North America there is little evidence that the virulence structure of *Puccinia graminis tritici* is directly determined by the resistance genes of the wheat fields, except possibly in the south or in relation to the gene *SrTmp* (see Section 4.6). In eastern Australia *P. graminis tritici* has almost invariably adapted itself to the wheat fields through appropriate changes from avirulence to virulence for any new *Sr* genes the wheat breeders introduce. To control stem rust, wheat breeders have had to keep a step ahead of the pathogen, and breeding for rust resistance has been a never-ending process. Australia's experience with wheat stem rust, much more than North America's, is relevant to the central difficulty in controlling plant disease by vertical resistance: The pathogen adapts, and resistance is "lost."

Ecologically the wheat fields of North American and Australia differ widely. In North America east of the Rocky Mountains great differences in temperature between summer and winter inhibit *P. graminis tritici* from oversummering or overwintering locally. Except in the south, inocula must travel far, often from one *Sr* genotype to another. Adaptation is impeded. In Australia seasonal variation is less. *P. graminis tritici* can survive locally on volunteer wheat plants and uncultivated grasses when it is not in the wheat fields themselves. Adaptation is promoted.

5.2   DIRECT ADAPTATION OF THE PATHOGEN TO THE HOST

The first deliberate introduction of stem rust resistance genes (*Sr* genes) into Australian wheat fields was in 1938. The cultivar Eureka was released in eastern Australia; it was the first popular stem rust resistant wheat originating from a breeding program in Australia. It had the gene *Sr6*. At the time of its release Eureka seemed resistant to the whole Australian population of *P. graminis tritici*. Farmers welcomed it and planted it on an increasing scale, until a peak of acreage was reached in 1945. But in 1941, soon after Eureka's release, virulence for *Sr6* appeared in *P. graminis tritici*, and increasing virulence for *Sr6* kept pace with the increasing acreage of Eureka (see Fig. 5.1). After 1945 both the acreage of Eureka and the proportion of virulence for *Sr6* in the rust population declined. By 1957 virulence for *Sr6* was no longer found in surveys. This prompted a second spell of interest by farmers in Eureka. Again, virulence for *Sr6* appeared and kept pace with the increasing acreage of Eureka. Consequently, the renewed interest in Eureka proved to be fleeting. The pathogen had won.

In 1945, soon after Eureka became susceptible, a series of new cultivars with gene *Sr11* was released in Queensland and northern New South Wales. Gabo, Charter, Kendee, and Yalta were among them, and the acreage sown to them increased sharply. For 2 years the new cultivars remained free from stem rust, but in 1948 new virulent races appeared that attacked them. This put an end to the sharp rise in popularity (see Fig. 5.2), though the cultivars continued to be planted despite their susceptibility.

The introduction of cultivars with gene *Sr11* was followed by the introduction of cultivars with *Sr9b* (Gamenya and Festival), *Sr17* (Spica), and *SrTt1*

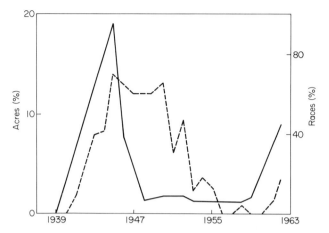

Fig. 5.1.   Virulent races of *Puccinia graminis tritici* in relation to the wheat variety Eureka which has the gene *Sr6*. The unbroken line shows the acreage of Eureka in northern New South Wales expressed as a percentage of the total wheat acreage. The broken line shows the frequency in northern New South Wales and Queensland of isolates of races of *P. graminis tritici* able to attack Eureka, expressed as a percentage of all isolates identified. From Vanderplank (1968, p. 67); data of Watson and Luig (1963).

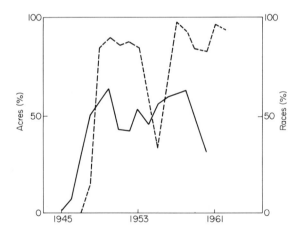

Fig. 5.2.   Virulent races of *Puccinia graminis tritici* in relation to the wheat variety Gabo and other varieties with the gene *Sr11*. The unbroken line shows the acreage of these varieties in northern New South Wales expressed as a percentage of total wheat acreage. The broken line shows the frequency in northern New South Wales and Queensland of isolates of races of *P. graminis tritici* able to attack Gabo, expressed as a percentage of all isolates identified. From Vanderplank (1968, p. 33); data of Watson and Luig (1963).

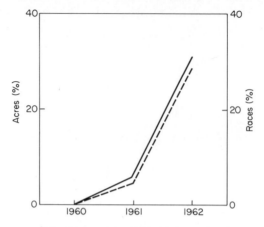

Fig. 5.3. Virulent races of *Puccinia graminis tritici* in relation to the wheat variety Mengavi which has the gene *SrTt1*. The unbroken line shows the acreage of Mengavi in Queensland expressed as a percentage of the total wheat acreage. The broken line shows the frequency in northern New South Wales of isolates of races of *P. graminis tritici* able to attack Mengavi, expressed as a percentage of all isolates identified. From Vanderplank (1968, p. 34); data of Watson and Luig (1963).

(Mengavi). The sequence of events was repeated, with new virulence following new resistance, but in detail the pattern varied. The genes *Sr9b* and *Sr17* were matched by virulence more slowly and *SrTt1* more quickly than genes *Sr6* and *Sr11*. Figure 5.3 illustrates the very quick matching of gene *SrTt1*.

After their experience with the transience of resistance given by *Sr6*, *Sr11*,

TABLE 5.1

**Percentage of Isolates of *Puccinia graminis tritici*
in Northern New South Wales and Queensland,
Arranged according to the Number of Stem Rust
Resistance Genes for Which They Were Virulent**[a]

Virulent for	Period		
	1954–1958	1959–1963	1964–1968
0 genes	8.6	0.0	0.0
1 gene	36.6	0.9	0.7
2 genes	54.7	46.2	6.6
3 genes	0.1	48.0	73.4
4 genes	0.0	4.9	12.5
5 genes	0.0	0.0	6.7
6 genes	0.0	0.0	0.0

[a] Data of Luig and Watson (1970).

TABLE 5.2

Percentage of Isolates of *Puccinia graminis tritici*
in Southern New South Wales, Arranged according
to the Number of Resistance Genes
for Which They Were Virulent[a]

Virulent for	Period		
	1954–1958	1959–1963	1964–1968
0 genes	15.2	0.2	0.0
1 gene	51.7	7.5	1.4
2 genes	33.1	73.9	46.4
3 genes	0.0	16.0	50.3
4 genes	0.0	2.4	1.0
5 genes	0.0	0.0	0.9
6 genes	0.0	0.0	0.0

[a] Data of Luig and Watson (1970).

Sr9b, *Sr17*, and *SrTt1* introduced singly, wheat breeders turned to accumulating resistance genes to provide a broader resistance base. Genes *Sr2, 5, 6, 7a, 8, 9b, 9g, 11, 12, 13, 15, 17, 26, 30*, and *Tt1* were all used, several cultivars combining four or five of them, e.g., Gatcher having *Sr5, 6, 8, 9g*, and *12*. *Puccinia graminis tritici* responded to a broader resistance by a broader virulence. This is shown by Tables 5.1 and 5.2, for the northern and the southern section of the eastern Australian wheat belt, respectively. By 1979/1980 the most abundant race of *P. graminis tritici*, race 343-1, 2, 3, 5, 6, in eastern and southern Australia was virulent for five genes, *Sr6, 8, 9b, 11* and *17*; it caused stem rust epidemics locally. *Puccinia graminis tritici* has an ability to adapt, the limit to which has not yet been probed, and stem rust in Australian wheat has been curbed only by the wheat breeders' continuous efforts.

5.3 INDIRECT ADAPTATION OF THE PATHOGEN TO THE HOST

As an illustration of indirect adaptation, virulence for the gene *Sr7b* was universal in *P. graminis tritici* in eastern Australia 25 years ago, but has now dwindled to comparative rarity. This original abundance and subsequent rarity have come about despite the fact that Australian wheat breeders have not used gene *Sr7b*. Virulence for gene *Sr7b* changed to avirulence in the continuous absence of gene *Sr7b* from Australian wheat cultivars. The change came about as an indirect adaptation to the wheat breeders' using genes *Sr6, 9b, 15*, and *17*.

Indirect adaptation results from the ABC–XYZ grouping of genes. In Australia evidence for indirect adaptation is somewhat obscured by direct adaptation. In the pathogen external adaptation to the host, i.e., direct adaptation, vies with the interallelic and intraallelic interactions that determine indirect adaptation; and in Australia external adaptation is strong. The pattern of necessary virulence, which allows the pathogen to attack the host, obscures the pattern of unnecessary virulence, which exists in the absence of the corresponding resistance genes.

In Australia the evidence for indirect adaptation is further obscured by the *Sr* genes of the XYZ group that breeders have chosen to use. The most widely used of them, gene *Sr11*, has only a feeble membership of the XYZ group. In the United States this feebleness is apparent in Table 4.2. In Canada, relative to gene *Sr6* (the relation most pertinent to Australian data), gene *Sr11* behaved neutrally, without evidence for membership of the XYZ group, in 1974, 1975, 1977, and 1978 (see Table 4.4). The other gene of the XYZ group, gene *SrTt1*, used by wheat breeders in Australia, is less feebly a member but still not a strong member of this group. By contrast with the XYZ group, the ABC group is strongly represented in Australia, with genes *Sr6*, *Sr9b*, *Sr15* and *Sr17* used by wheat breeders.

It will be remembered from the previous chapter that unnecessary virulence for genes, one from the ABC group and one from the XYZ group, tends to dissociate, with combined virulence less than what one would expect if virulence were randomly distributed, whereas virulence for genes both of the ABC or both of the XYZ groups tends to associate, with combined virulence more than one would expect from virulence randomly distributed.

The years of Australian wheat breeding can be divided. In the early years virulence for the XYZ group preponderated, in the latter years virulence for the ABC group. The early years, from 1952 through 1963, were surveyed by Watson and Luig (1963). Information about the latter years was given by Luig (1980, 1981) and R. A. McIntosh (personal communication). In the early years virulence for gene *Sr7b* was universal (it had been in the wild-type population of *P. graminis tritici* before the first stem rust resistant wheat variety was introduced in Australia) and virulence for gene *Sr11* was common (see Fig. 5.2). In the latter years, after many wheat cultivars with genes *Sr6*, *Sr9b*, *Sr15*, and *Sr17* had been introduced, the population of *P. graminis tritici* changed, and virulence for these *Sr* genes was common. Virulence for genes *Sr6*, *Sr9b*, and *Sr17* was closely associated and predominant.

Consider first virulence for the XYZ group. The early years encouraged it, by the association of virulences. When the gene *SrTt1* was introduced in Australia, the response of the pathogen was immediate; virulence developed swiftly (see Fig. 5.3). There is evidence that the swift development of virulence was helped by virulence association within the XYZ group. Of the 336 isolates virulent for *SrTt1*, 333 or 99% were also virulent for genes *Sr7b* and *Sr11*, i.e., there was associated virulence for genes *Sr7b*, *Sr11*, and *SrTt1*, all of the XYZ group. The remaining 3 isolates had associated virulences for *Sr7b* and *SrTt1* but not *Sr11*. Dissociation from virulence for genes of the ABC group was also apparent. Of the 336 isolates only 1 was also virulent for gene *Sr6*. This is disproportionately low and significantly less than expected; of the total 5322 isolates analyzed during this period, 399 were virulent for gene *Sr6*. (Significant dissociation of virulence for gene *SrTt1* from virulence for gene *Sr6* is also apparent if one confines the analysis to the relevant years, 1960 through 1963, and relevant regions, New South Wales and Queensland.) There was total dissociation of virulence for gene *SrTt1* from virulence for genes *Sr9b* and *Sr17*, a dissociation that is significant statistically. Virulence for these two genes of the ABC group had already become quite common in the general population of *P. graminis tritici*; of the total 5322 isolates, 431 were virulent for gene *Sr9b* and 312 for gene *Sr17*.

When virulence for the XYZ group is followed into the latter years that have an ABC background, the effects of virulence dissociation become even clearer. In these latter years virulence for gene *SrTt1* practically disappeared; in 1979/1980 none was found, although gene *SrTt1* itself is present in several wheat cultivars. Virulence for *Sr7b* was greatly reduced and, except in Western Australia, was absent from the most widely distributed stem rust races. Only virulence for gene *Sr11* remained prevalent, almost universal, perhaps because of the weakness of its membership of the XYZ group and because gene *Sr11* had been used by wheat breeders for so long.

The story of virulence for the ABC group is the opposite to that of virulence for the XYZ group. Unnecessary virulence for the ABC group was depressed in the early years. Figure 5.1 shows how quickly virulence for gene *Sr6* was reduced as soon as it became unnecessary. Necessary virulence for genes *Sr9b* and *Sr17* developed after the introduction of cultivars with these genes, but the development was much slower than that of virulence for *SrTt1*.

Table 5.3 gives a background to Fig. 5.1. For what artifacts are worth, it shows how the combination of virulence for gene *Sr6* of the ABC group with virulence for genes *Sr7b* and *11* of the XYZ group reduces the fitness of *P. graminis tritici* to survive on a wheat variety equally susceptible to both virulence combinations.

The change in the latter years to preponderant virulence for the ABC group of genes is obviously a direct adaptation. Wheat breeders used the genes *Sr6*, *Sr9b*, *Sr15*, and *Sr17*, and the fungus adapted accordingly. The association of virulence for genes *Sr6*, *Sr9b*, and *Sr17* in the prevalent races could also be a direct adaptation, although these genes are not together in the common wheat cultivars. Nevertheless, if one can believe the evidence of

TABLE 5.3

**Effect of Passage through the Susceptible Wheat Variety Federation
on the Percentage of Two Virulence Combinations
in *Puccinia graminis tritici*[a]**

		No. of passages through Federation			
Test[b]	Virulence for	1	3	4	5
6	*Sr7b* and *11*[c]	34.4	95.3	91.7	94.8
	Sr7b, 11, and *6*[d]	65.6	4.7	8.3	5.2
9	*Sr7b* and *11*[c]	41.1	84.5	92.2	96.1
	Sr7b, 11, and *6*[d]	58.9	15.5	7.8	3.9

[a] Data of Watson and Singh (1952).

[b] Tests 6 and 9 refer to Watson and Singh's mixtures 6 and 9. Data for an irrelevant race are excluded from test 9, and the percentages adjusted accordingly.

[c] Race 222–2,6.

[d] Race 222–1,2,6.

artifacts, indirect adaptation might also have been involved. Table 5.4 uses Watson and Luig's (1968a) data to compare the fitness in a susceptible wheat variety of two combinations of virulence in *P. graminis tritici*. The one combination was fairly well balanced, with virulence for gene *Sr7b* from the XYZ group and for gene *Sr15* from the ABC group. The other combination was unbalanced by the addition of virulence for genes *Sr6* and *Sr9b*, both strong members of the ABC group, and for gene *Sr11*, a weak, nearly neutral member of the XYZ group. Table 5.4 shows that this combination, with an unbalanced excess of virulence for genes of the ABC group, steadily increased with each passage through the susceptible wheat host variety. This combination partially mimics the prevalent virulences in eastern Australia in the latter years, with virulence for gene *Sr15* substituted for virulence for gene *Sr17*, also of the ABC group, and with virulence instead of avirulence for *Sr7b*. Unbalanced virulence, even if the virulence is not directly necessary for parasitism, can increase the parasite's fitness.

It is part of the concept of the ABC–XYZ grouping of genes that virulences for genes within a group tend to associate. Paraphrased, virulences within a group tend to increase the contribution to the pathogen's fitness of other virulences within the same group, which is what Table 5.4 suggests. All this carries a warning. The plant breeder's procedure of accumulating resistance genes within cultivars to meet the threat from the pathogen (see Tables 5.1 and 5.2) is not without its danger. Imbalance between the ABC and the XYZ groups of genes during the process of accumulating genes could indirectly

TABLE 5.4

Effect of Passage through a Susceptible Wheat Variety on the Percentage
of Two Virulence Combinations in *Puccinia graminis tritici* [a]

Test	Virulence for	No. of passages			
		1	2	3	4
1	*Sr7b* and *15* [b]	22	15	7	3
	Sr7b, 11, 6, 9b, and *15* [c]	78	85	93	97
2	*Sr7b* and *15*	42	33	15	7
	Sr7b, 11, 6, 9b, and *15*	58	67	85	93
3	*Sr7b* and *15*	30	16	7	2
	Sr7b, 11, 6, 9b, and *15*	70	84	93	98

[a] From data of Watson and Luig (1968a). The tests were duplicated, and the duplicates agreed closely.
[b] Race 21-7.
[c] Race 21-1, 2, 3, 7.

help the pathogen. Gene balance, it would seem, is as important as gene number in the plant breeder's procedure.

5.4   DISCUSSION

Success in controlling plant disease by vertical resistance using pseudo-monogenes depends on two balances: the balance between direct and indirect adaptation of the pathogen to the host, and the balance between genes of the ABC and XYZ groups.

Vertical resistance is more likely to be successful when for ecological reasons direct adaptation of the pathogen to the host is restricted, as it is with stem rust of wheat in the north of North America. It is almost certain to be unsuccessful when direct adaptation is favored, as it usually is with diseases of perennial crops or with most diseases in the tropics and subtropics where host plants are found with varying abundance all the year round. The Australian data on wheat stem rust are useful because they reflect an intermediate ecology.

Little is known of the ABC–XYZ groups in diseases other than wheat stem rust, and even the *Sr* genes have been very incompletely analyzed. We lack knowledge of the grouping of most genes, and even when we know the grouping we do not know exactly how to exploit it. Balance can be obtained by using one gene from each group, e.g., in wheat stem rust *Sr6* from the

ABC group and *Sr9e* from the XYZ group. Would it help to add a second gene from each group? Or would the tendency for virulence to associate when it is for genes within a group outweigh any advantage given by having more than one gene of a group? These are questions that cannot yet be answered; until they are answered, the application of pseudomonogenes for disease resistance will involve much floundering.

6

Mutation in the Pathogen from Avirulence to Virulence

6.1 VARIABLE MUTATION RATES

Mutation, the ultimate means of adaptation of pathogen to host, occurs at a variable rate.

Avirulence is usually dominant over virulence, and loci heterozygous for avirulence are more easily mutated to phenotypic virulence than homozygous loci (Flor, 1960). The rate varies with the locus. Flor (1958) used an F_1 hybrid of *Melampsora lini* heterozygous for avirulence at four loci and found significant differences in mutation rates. At one locus two natural mutants were found in 200,000 uredospores; at another, one natural mutant in 600,000 uredospores; and at the other two, no mutants in 300,000 and 900,000 uredospores, respectively.

Luig (1979) studied mutation in *Puccinia graminis tritici* in natural conditions and also with a mutagen (ethyl methanesulfonate). At the one extreme, avirulence for the resistance genes *Sr5*, *Sr15*, *Sr21*, and *Sr9e* in wheat was found to have a very high spontaneous mutation rate to virulence, as well as a high rate after treatment with the mutagen. At the other extreme is the stability of avirulence for the gene *Sr26* derived from *Agropyron elongatum*. This gene is in the Australian wheat cultivar Eagle released in 1971 and in the cultivars Kite and Jabiru. The popularity of these cultivars increased swiftly, and by 1974 Eagle covered 530,000 hectares and by 1975 it covered 655,000 hectares in Australia. All efforts by agronomists to find stem-rusted plants of these cultivars failed. Also, because it was considered unwise to rely on monogenic resistance, a great effort was made to find a virulent mutant that could be used experimentally by wheat breeders in order to enable them to incorporate other resistance genes. Mutagens were used, all to no avail. Whatever the end of the story may be, the evidence will stand that mutation from avirulence to virulence for the gene *Sr26* was rare after it was introduced.

Luig suggested that *P. graminis tritici* possesses intrinsically different genes for avirulence which he grouped into five classes: (1) genes like those conditioning avirulence for the genes *Sr5*, *Sr9e*, and *Sr21*, which mutate readily to virulence both in the field and after being subjected to mutagen treatment; (2) genes like that conditioning avirulence for the gene *SrTt1*, which have a substantial spontaneous mutation rate which was not increased by treatment with ethyl methanesulfonate; (3) genes like those conditioning avirulence for the genes *Sr6* and *Sr30*, which have a low mutation rate both under natural conditions and after being treated with a mutagen; (4) genes like those conditioning avirulence for the genes *Sr13*, *Sr24* (from *Agropyron elongatum*), and *Sr27* (from rye), which rarely mutate to virulence; and (5) the gene that conditions avirulence for the gene *Sr26*, which in 1979 had not been known to mutate to virulence.

6.2   WILD-TYPE PATHOGEN POPULATIONS

For the purpose of discussion a wild-type population is taken to mean the population of the pathogen as it existed on the host before resistance genes were introduced or before some particular resistance gene was introduced.

If the pathogen population is large enough, virulence preexists. It is there before the resistance gene is introduced into the host plants, and a boom-and-bust cycle involves the multiplication of virulence already present.

If the virulence allele were slightly less fit to survive than the avirulence allele, its frequency in the wild-type population would be roughly of the same order as the mutation rate. With a mutation rate of, say, 10^{-7} the frequency would be of the order of 10^{-7}. The frequency of double mutants, i.e., of mutation to virulence for two R genes would be of the order of 10^{-14} provided that there was no epistasis. If one of the R genes belonged to an ABC group and the other to an XYZ group the frequency of the double mutant would be less than 10^{-14}. If both genes belonged to an ABC group or both to an XYZ group, the frequency would be greater than 10^{-14}. Without epistasis the frequency of a triple mutant would be of the order of 10^{-21}; and so on.

If the virulence allele were fitter than the avirulence allele, its frequency in the wild-type population would exceed the mutation rate. If it were consistently fitter, the virulence allele would be taken up in the population; we would be unlikely to know of it or the corresponding R gene in the host. If it were inconsistently fitter, the virulence allele would be frequent but known. Virulence in *Phytophthora infestans* for the potato gene $R4$ was frequent in the wild-type population. Consequently, no resistant potato variety with the gene $R4$ alone was bred, although the gene $R4$ occurs in a few varieties in combination with other R genes.

Reports that relevant virulence was absent in the wild-type population mean that the sample of the population was too small to reveal the virulence's presence. In plant breeding practice the most common spore trap is the experimental plot or field. New cultivars are released if they are resistant in these plots or fields. After release, the cultivated area expands until eventually the trap is large enough to catch virulent spores. With vertical resistance it is the safest policy to assume that infections will sooner or later occur. (Even then the virulent mutants may escape notice and infection may be reported as nil because, although the cultivated area may be large, the area actually examined is usually limited, and single infections here and there are not necessarily conspicuous.) Whether the resistance will then be stable or not depends on whether the infections fail or succeed in reproducing themselves. Selection pressures and not mutation determine the stability of resistance ultimately, but mutation rates determine how long the change will be postponed.

6.3   VIRAL DISEASES

Most plant viruses have single-stranded RNA which seems to be highly mutable (Hennig and Wittmann, 1972). In viruses with multipartite genomes recombination can produce much variation from few mutations. Nevertheless, virus resistance in agriculture has proved to be rather stable, with some exceptions.

Tomato spotted wilt is a virus existing as a number of strains under different genetic control (Finlay, 1952). Tomato varieties resistant in the Hawaiian Islands and New Jersey were found to be susceptible in Western Australia. Yet despite this variability there are few records of tomato varieties becoming susceptible where they were previously resistant: this stability is perhaps the result of spotted wilt being essentially exodemic in tomatoes.

Similarly, barley yellow dwarf virus exists in different strains. Ethiopian barleys resistant to this virus in some parts of North America are susceptible elsewhere, and in Canada, Gill and Buchannon (1972) found a differential reaction between four isolates of the virus and barley genotypes.

Resistance in tomato to tobacco mosaic virus is also unstable. Pelham *et al.* (1970) and Pelham (1972) demonstrated a strong influence of the host on the pathogen in Britain. Prior to 1966 only avirulent strains were found. In 1966 resistant tomato varieties, with the gene *Tm1*, were introduced. By 1967 the frequency of virulent strains in greenhouses growing these resistant varieties had increased to 50%, and by 1968 to 93%. Stabilizing selection in favor of avirulence is equally strong. When resistant varieties were again replaced by susceptible varieties, virulent strains soon decreased in prevalence.

Resistance in raspberry to raspberry ringspot virus is unstable, but stabilizing selection against virulence is strong. Virulent strains are less frequently transmitted through the seed (Murant *et al.*, 1968), and seed transmissibility and virulence are controlled by the same part of the virus genome (Hanada and Harrison, 1977).

The virus genome is small, and it is perhaps to be expected that a mutation to virulence would have relatively large side effects on parasitic performance. It may well be a rule that the strength of stabilizing selection in favor of wild-type populations is inversely proportional to the size of the genome, other things being equal.

6.4   BACTERIAL DISEASES

There seems to be no generally accepted figure for the mutation rate from avirulence to virulence in phytopathogenic bacteria, but it is known that bacteria have efficient processes for recombining mutant DNA. Virulence may be transferred by plasmids, and genes may be reassorted by transduction, transformation, and sexduction.

In bacterial diseases evidence that the introduction of resistant host plants brings about a change toward more frequent virulence in the pathogen has been noted for blight of cotton caused by *Xanthomonas malvacearum*. Bacterial blight is the major disease of cotton in Africa and elsewhere where wind-driven rain or overhead irrigation disperses the pathogen. There are various control measures, among them the use of resistant varieties. Breeding cotton for resistance to *X. malvacearum* has been a major project in Africa for half a century and is a common project elsewhere. The literature was reviewed by Brinkerhoff (1970). When resistance was conditioned by few genes it was often unstable. Thus, in the United States the resistance of cotton varieties with single *B* genes in a susceptible background was matched in a single season by increased virulence in *X. malvacearum*. Indeed, Brinkerhoff (1963) found that inoculum from Uganda was virulent for all the genes *B1*, *B2*, *B3*, *B4*, *B5*, *B7*, *Bin*, and *Bn*. Cross (1963) in East Africa found that the resistant variety Albar 51 and its derivatives were matched by appropriate cultures of the bacterium. Nevertheless, cotton breeders seem to have succeeded in synthesizing high levels of resistance which appears to be stable.

Bacterial blight of rice caused by *X. oryzae* is found in most rice-growing areas of Asia and has recently been reported in America. It is a highly variable organism, and breeding for resistance has been undertaken in many countries. There have been reports of the breakdown of resistance (Rao *et al.*, 1971; Reddy and Ou, 1976), but these have been surprisingly few, considering the variability of the pathogen.

A destructive adaptation of bacteria to the host was reported by Crosse (1975) in England. Before 1971 the variety Roundel of sweet cherry was substantially resistant to *Pseudomonas mors-prunorum*, but in 1971 a variant of this bacterium arose which differed pathogenically from other strains and was highly destructive on Roundel.

6.5 FUNGAL DISEASES

The earliest report of mutation from avirulence to virulence seems to be that of Gassner and Straib (1932) in *Puccinia striiformis*. In a single-spore culture and its progeny, mutation to virulence was observed at the rate of 1 in 60,000–120,000. The mutant form was able to attack many wheat varieties resistant to the original culture. The mutant was found in 34 cultures and was first observed in greenhouse experiments. Later it was found in experimental plots of wheat. Another early report was that of Newton and Johnson (1939) who observed a mutation from avirulence to virulence in a culture of *P. graminis tritici*. The record is unusual because it is for mutation during the storage of uredospores in a refrigerator. The possibility of contamination was excluded. Since then, reports of mutation have been frequent. Only two pathogens will be considered in detail.

First consider *Phytophthora infestans* in potatoes. Frequent mutation has been demonstrated. Eide *et al.* (1959) used cultures derived from single spores of race (0), which is avirulent for all known *R* genes, and observed frequent mutation to virulence for the genes *R2*, *R3*, and *R4*. Black (1960) recorded mutation from avirulence to virulence for the gene *R1* and, very frequently, *R4*. Graham *et al.* (1961) recorded mutation from avirulence to virulence for the genes *R3* and *R5*, Denward (1970) for the genes *R3* and *R4*, and Shattock (1976) for the gene *R2*. Earlier, Howatt and Grainger

(1955) in a greenhouse inoculation experiment found spectacular evidence for the mutability of *P. infestans*. They established potato plants in pots in a special chamber within a greenhouse. The temperature was controlled so that it never exceeded 21°C. Spray nozzles humidified the chamber every hour, adding the moisture equivalent to 18 mm of rainfall per day. Into the chamber they introduced 10,000 seedlings of *Solanum demissum* × *Solanum tuberosum* crosses. Inoculum of race (0) was introduced, and plants without *R* genes became infected. Some weeks after this it was noticed that blight was spreading to plants not previously attacked. Eventually, all the seedlings were blighted, and isolations showed that the fungus now belonged to many races, including some virulent for all four resistance genes, *R1*, *R2*, *R3*, and *R4*, which were then known. Mutation followed by selection in a mixed host population under almost ideal environmental conditions for blight allowed a change to occur from race (0) to race (1, 2, 3, 4) within a few weeks. The ideal conditions did not promote the mutation, but they did promote the survival and therefore the identification of the mutants. There is evidence for combination in *P. infestans*. Malcolmson (1970) mixed sporangia of race (3, 4, 10) with sporangia of race (1, 2, 3, 7). After several cycles of inoculation, race (1, 2, 3, 4, 7, 10) was recovered.

Mutation in *Erysiphe graminis tritici* was studied by Leijerstam (1972) in the greenhouse. He grew seedlings of 10 mildew-resistant wheat varieties and 1 susceptible variety, Thatcher, in pots, with precautions against unintended contamination. Each pot contained one resistant and one Thatcher seedling. The seedlings were inoculated with spores from a race of *E. graminis tritici* avirulent for all the resistant varieties, in such a way as to produce a uniform infection covering about 50% of the leaf area of Thatcher. When the Thatcher plants began to die from mildew a new set of seedlings was inoculated. In populations maintained for 438 days three mutants appeared. One mutant was virulent for the gene *Pm1*, another for *Pm2*, and a third for *Pm3a*. A fourth mutant appeared in a population maintained for 241 days. It was virulent for three resistance genes: *Pm1*, *Pm4*, and *Mle*. (It would seem on this evidence that there was a tendency for virulence association, i.e., that genes *Pm1*, *Pm4*, and *Mle* all belong to an ABC group or all to an XYZ group.) On the assumption that there would be 5 million wheat plants per hectare and that 5% of the leaf surface would be mildewed, Leijerstam estimated that there would be 1141 mutations per hectare per day for the three single loci and 2074 for the triple mutant.

These mutation rates for *Phytophthora infestans* and *E. graminis* are very high. Were they operative in the field as well as the greenhouse, there would have been little usable vertical resistance to disease in agriculture. But they are not the operative rates in the field.

6.6   INOPERATIVE AND OPERATIVE MUTATION

Mutation from avirulence to virulence is unusual in biology in that the mutation is initially inoperative. Mutation occurs in populations of the pathogen; populations of the pathogen occur on susceptible varieties of the host; and virulence does not operate differently from avirulence in varieties susceptible to the avirulence allele. Agriculturally, a mutation from avirulence to virulence becomes significant only when the mutant reaches a field with the corresponding R gene.

If there is no strong stabilizing selection against virulence in the wild-type population, the frequency of mutants is approximately the genetic mutation rate (see Section 6.2). The probability that a spore (or bacterial cell or virus particle) will reach a field with the appropriate R gene is a factor in the operative mutation rate. Thus

$$u_O \sim u_G \times p$$

where u_O is the operative mutation rate, u_G the genetic mutation rate, and p is the probability that a spore produced in the wild-type population will reach a resistant plant. If there is stabilizing selection against virulence in the wild-type population,

$$u_O < u_G \times p$$

The factor p is highly variable and depends among other things on the area of cultivation of the resistant variety. This is a matter taken up in a different context in Chapter 13.

Because p is often very small, it follows that the operative mutation rate is often very much less than mutation rates measured in the laboratory or greenhouse.

6.7   EPIDEMIOLOGICAL MUTATION

Operative mutation is perhaps best considered along with epidemiological mutation, which is a facet of population genetics. Epidemiological mutations from avirulence to virulence are changes in the parasite population as it occurs on the host population. To take an example from Section 6.1 phenotypic mutation occurs readily from avirulence to virulence for the genes *Sr9e* and *Sr21*; but Luig (1979) observed that these mutations arise frequently in the Australian rust flora only to die out in a season or two without the virulent form persisting. In the particular conditions observed by Luig and in the particular years of his observations, phenotypic mutation from avirulence to virulence for the genes *Sr9e* and *Sr21* was not followed by epidemiological mutation; in relation to virulence for the genes *Sr9e* and *Sr21* the Australian population of *Puccinia graminis tritici* remained for all practical purposes unmutated. In general it would seem that epidemiological mutation depends more on selection than on genetic or phenotypic mutation. Thus, virulence for the gene *Sr30* is very rare in isolates of *P. graminis tritici* in North America but almost universal in eastern Australia (Luig, 1979), where there was an epidemiological mutation from rarity in the early 1960s to abundance now.

Of the sorts of mutation, epidemiological mutation is the only sort in which the frequency of mutation per unit of pathogen, e.g., per million spores, depends on the area of cultivation of the host. One can picture the pathogen population in a state of recurrent genetic and phenotypic mutation, with recurrent and local flares of virulence in different loci. Some of the flares catch on to become the conflagration we call epidemics. Most of the flares simply die out. This statement is essentially an adaptation of Fisher's (1930) postulate that mutations are recurrent, and any mutations now existing must have occurred many times before in the history of the species. To survive, for the flare to become a conflagration, the pathogen must increase, with progeny exceeding parents in number year by year. Its best chance is when cultivation is extensive.

Epidemiological mutation involves the pathogen on resistant varieties, because the mutation would have no epidemiological significance unless it allowed previously resistant varieties to be attacked. For genetic mutants to become epidemiological mutants they must survive first of all in the susceptible crop where they originated; then they must reach a resistant crop, e.g., by wind-blown spores; and in the resistant crop they must survive from year to year, with all the hazards of survival in the off-season if the crop plants are annuals. The chance of reaching a resistant crop and of surviving from year to year increases with the area under cultivation. This is the host effect in breeding for resistance to disease. A successful breeding program is one that introduces a new resistant cultivar over a large area, giving the pathogen great scope for survival, and does so without incurring epidemiological mutation.

7

Horizontal and
Vertical Resistance

Horizontal and vertical resistance are defined in terms of only two variables: host and pathogen.

In such a two-variable system, resistance of the host to disease must exist in two, and no more than two, forms. The main effect of variation of the host, using the term "main effect" strictly in its biometric sense, determines horizontal resistance. Interaction, i.e., the differential effect, determines vertical resistance. These are definitions, pure and simple, illustrated biometrically. In these biometric definitions in a two-variable system, horizontal and vertical resistance, together or separately, describe every possible form of resistance.

So, too, in a two-variable system pathogenicity must exist in two, and no more than two, forms. The main effect of variation in the pathogen, again using the term "main effect" strictly in its biometric sense, determines aggressiveness, and the interaction, i.e., the differential effect, virulence. These are definitions, illustrated biometrically. In these biometric definitions in a two-variable system, aggressiveness and virulence, together or separately, describe every possible form of pathogenicity.

To put the matter in another way for clarity, horizontal and vertical resistance and aggressiveness and virulence are precisely defined terms to cover all possibilities within a two-variable system of host and pathogen.

TABLE 7.1

**Analysis of Variance in Relation to Definitions of Horizontal Resistance
and Vertical Resistance in the Host, and Aggressiveness and Virulence
in the Pathogen in a Two-Variable System**

Main effect between host varieties	Horizontal resistance
Main effect between pathogen isolates	Aggressiveness
Interaction varieties × isolates	Vertical resistance, and virulence

Horizontal resistance and vertical resistance can, and often do, coexist. There is no either/or condition. Resistance in any one host plant may be a mixture of horizontal and vertical, in any proportion, including 0 and 100%, because main effects and interactions, as biometric concepts, may occur together, in any proportion.

So, too, aggressiveness and virulence can and often do coexist without an either/or condition.

Table 7.1 summarizes the definitions of the two sorts of resistance and of pathogenicity. No categories are possible in a two-variable system other than main effects (horizontal resistance and aggressiveness) and interactions (vertical resistance and virulence). These are definitions in terms of variation. The sources of variation fall outside the definitions. There is no limit to the sorts of biochemical, physiological, morphological, or anatomical variation that contribute to the total variation. There might be, e.g., many different biochemical pathways involved in horizontal resistance or vertical resistance; the definitions do not preclude this.

In these definitions, irrespective of whether they are for horizontal or vertical resistance, the terms "more resistant" and "less susceptible" are equivalent and interchangeable, their aptness varying only with the background. The point demonstrated in Chapter 2, that specificity resides in susceptibility, is therefore irrelevant to the definitions and therefore to the present chapter.

7.2   THE GEOMETRIC ILLUSTRATION

Vanderplank (1963) used a geometric illustration for the two possible sorts of resistance a host plant can have in a two-variable system. Horizontal resistance is expressed uniformly against races of the pathogen, and vertical resistance is expressed differently.

Figure 7.1 shows the behavior of two potato varieties, Kennebec and Maritta, infected with *Phytophthora infestans*. Both these varieties have the resistance gene *R1*. This gene confers resistance to races (0), (2), (3), (4),

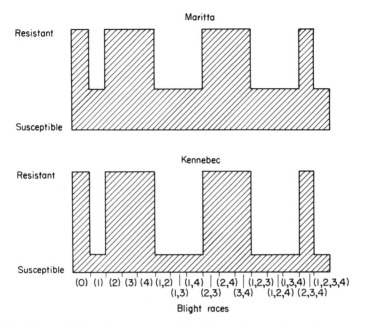

Fig. 7.1. Diagram of the resistance to blight of the foliage of two potato varieties, Kennebec and Maritta, both with the resistance gene *R1*. The resistance is shown shaded to 16 races of blight. To races (0), (2), (3), (4), (2, 3), (2, 4), (3, 4), and (2, 3, 4) the resistance of both varieties is vertical. To races (1), (1, 2), (1, 3), (1, 4), (1, 2, 3), (1, 2, 4), (1, 3, 4), and (1, 2, 3, 4) resistance is horizontal, small in Kennebec and greater in Maritta. From Vanderplank (1963, p. 175).

(2, 3), (2, 4), (3, 4), and (2, 3, 4). It does not confer resistance to races (1), (1, 2), (1, 3), (1, 4), (1, 2, 3), (1, 2, 4) (1, 3, 4), and (1, 2, 3, 4). The resistance conferred by the gene *R1* is expressed differentially against the races and is vertical.

Against the races to which the gene *R1* does not confer resistance, Kennebec and Maritta behave differently. Kennebec succumbs faster to blight than Maritta. Grown side by side, Kennebec is blighted brown while Maritta is still mainly green. Maritta has more "field resistance," to use a common term; Maritta has more "horizontal resistance," to use our term. This is illustrated in Fig. 7.1. Where there is no vertical resistance against a race, the horizontal resistance is shown higher for Maritta than for Kennebec.

The resistance of Maritta (or Kennebec) depends on the composition of the spore shower of *P. infestans*. As almost every survey shows, populations of *P. infestans* are mixtures of races. Against a mixture of races (0) and (1) or against any mixture of races with and without 1 in their designation, Maritta must simultaneously exhibit both horizontal and vertical resistance: horizontal resistance to the race components with 1 and vertical resistance to those without 1 in their designation. Resistance in one and the same Maritta plant at one and the same time would be both horizontal and vertical. By every conceivable agricultural or laboratory test, using *P. infestans* populations from almost every country of the world where potato blight is a menace, Maritta would be resistant both horizontally (by comparison with Kennebec) and vertically, irrespective of whether single plants or whole fields were tested.

Figure 7.1, taken unchanged from the original 1963 publication, shows that there is no either/or condition in the definitions of horizontal and vertical resistance. Both forms of resistance not only can, but also must, coexist when, as is almost universally the case, the pathogen comes as an appropriate mixture of races.

7.3   ILLUSTRATION BY ANALYSIS OF VARIANCE

Consider for illustration some data of Paxman (1963) on *P. infestans* in potato tubers. Using potato varieties without *R* genes, Paxman set out to determine whether isolates of *P. infestans* would become specially adapted to a variety if they were grown continuously on it. He obtained an isolate 30 RS from a naturally infected tuber of the variety Red Skin, another isolate 31KP from a naturally infected tuber of the variety Kerr's Pink, and an isolate 32KE from a naturally infected tuber of the variety King Edward. These he subcultured, each on its original variety, i.e., he subcultured isolate 30RS on tubers of Red Skin, isolate 31KP on tubers of Kerr's Pink, and isolate 32 KE on tubers of King Edward. By the time he came to make his tests there had been 90 cycles of subculturing; for example, isolate 30RS had been on Red Skin for 90 cycles plus the unknown number of cycles the fungus had been on Red Skin in the field before it was isolated. In his tests he used as his criterion the rate of spread of mycelium in tuber tissue. He measured the rate of spread of each of his three isolates not only in the variety of origin, e.g., 30RS in Red Skin, but also in the other two varieties, e.g., 30 RS in Kerr's Pink and King Edward as well. In addition he tested a fourth isolate (of unspecified origin, cultured on the variety Majestic) in the three test varieties, doing two separate tests. The analysis of variance is given in Table 7.2.

There was a highly significant main effect, i.e., there was a highly significant difference in horizontal resistance between varieties, with 2 degrees of freedom. There was a highly significant main effect, i.e., there was a highly significant difference in aggressiveness between isolates, with 3 degrees of freedom. But the interaction varieties × isolates with 6 degrees of freedom was insignificant. That is, with the particular varieties and isolates he used,

TABLE 7.2

Combined Analysis of Variance in Two Tests of Four Isolates
of *Phytophthora infestans* in Three Potato Varieties[a]

	Degrees of freedom	Mean square	P
Varieties	2	919.0	<.001
Isolates	3	563.7	<.001
Varieties × isolates	6	23.5	
Error	132	19.6	

[a] Data of Paxman (1963).

and with his particular measurements of mycelial spread, there was no evidence for vertical resistance or virulence. (It will be remembered that the varieties Paxman used had no *R* genes.)

Also using the technique of the analysis of variance Caten (1970) obtained results very similar to those of Paxman. In tubers Caten measured the surface growth of the fungus in 9 days along the tuber from the point of inoculation. There were highly significant main effects of cultivars (i.e., in horizontal resistance) and of isolates (i.e., in aggressiveness), but the interaction cultivars × isolates was insignificant. There was no evidence for vertical resistance and virulence. In leaves he measured the latent period, from inoculation to the first appearance of sporangia. Again there were highly significant main effects of cultivars and of isolates; but the interaction cultivars × isolates was insignificant in two of the three tests and barely significant in the third.

7.4   ILLUSTRATION BY RANKING ORDER

If host variety 1 is horizontally more resistant than 2 when challenged by race 1 of the pathogen, it will be more resistant than 2 when challenged by any other race. Put generally, with horizontal resistance variation in the population of the host, expressed by ranking the host variants in order of increasing or decreasing resistance, is independent of variation in the population of the pathogen. With vertical resistance, on the other hand, the ranking order of the host variants depends on which variant of the pathogen is involved.

Consider an example of uniform ranking characteristic of horizontal resistance. The data are those of Johnson and Law (1975) for stripe (yellow) rust in wheat caused by *Puccinia striiformis*. The cultivar Hybride de Bersée of hexaploid wheat is reputed to be moderately resistant to stripe rust. It has the normal 21 pairs of chromosomes and will be referred to as euploid

Bersée. The chromosomes carry a reciprocal translocation with respect to chromosomes 5B and 7B of the cultivar Chinese Spring. One chromosome of euploid Bersée corresponds with the long arms of chromosomes 5B and 7B in Chinese Spring and is therefore designated 5BL-7BL, whereas another of euploid Bersée corresponds with the short arms of these chromosomes in Chinese Spring and is designated 5BS-7BS. The second host genotype to be considered here is Hybride de Bersée monosomic 5BS-7BS, designated mono 5BS-7BS. It has 20 pairs of chromosomes together with a single short chromosome that corresponds with the short arms of chromosome 5B and 7B in Chinese Spring. The third genotype is Hybride de Bersée nullisomic 5BS-7BS, designated nulli 5BS-7BS. It has 20 pairs of chromosomes, lacking the pair 5BS-7BS. To turn from variants of the host to variants of the pathogen, five races of *P. striiformis*, listed in Table 7.3, were used to which all three host genotypes reacted after inoculation with a "3-" to "4-" type of lesion. Johnson and Law (1975) estimated the percentage of leaf area infected in the field. Table 7.3 gives mean estimates for two different dates of assessment, the three wheat types, and the five races of *P. striiformis*.

The data in Table 7.3 are reworked in Table 7.4 to rank the host types in order of decreasing resistance to the various races, and in Table 7.5 to rank the races in order of increasing aggressiveness on the various host types. The ranking order is uniform. Euploid Bersée was the most resistant host type to the races, and nulli 5B-7BS was the least resistant. So, too, race 40E136 was the least aggressive and race 40E8 the most aggressive, irrespective of the host type concerned; the other races fall into place. The uniformity of ranking shows that the resistance was substantially horizontal.

Consider now reversed ranking characteristic of vertical resistance. Table 7.6 illustrates reversed ranking in the simplest possible way. The *R1* genotype has resistance to race 2 but not race 1; the *R2* genotype has resistance to

TABLE 7.3

Present Leaf Area Diseased When Three Wheat Host Genotypes Were Infected with Cultures of Five Races of *Puccinia striiformis*[a]

| Host | Race | | | | |
	41E136	104E137	104E9	108E9	40E8
Euploid Bersée	0	0	<1	<1	2.8
Mono 5BS-7BS	1.0	1.3	1.3	2.5	15.0
Nulli 5BS-7BS	1.0	5.8	8.8	20.0	25.0

[a] Means of two plots and two dates, 27 June and 7 July, from data of a field test by Johnson and Law (1975).

TABLE 7.4

Three Wheat Host Genotypes Ranked in Order of Decreasing Resistance to Five Races of *Puccinia striiformis*[a]

	Race				
Host	41E136	104E137	104E9	108E9	40E8
Euploid Bersée	1	1	1	1	1
Mono 5BS-7BS	2.5	2	2	2	2
Nulli 5BS-7BS	2.5	3	3	3	3

[a] Data of Table 7.3, reworked; 1 = most resistant, 3 = least resistant, and 2.5 means a tie for second and third place.

TABLE 7.5

Five Races of *Puccinia striiformis* Ranked in Order of Increasing Aggressiveness to Three Wheat Host Genotypes[a]

	Host		
Race	Euploid Bersée	Mono 5BS-7BS	Nulli 5BS-7BS
41E136	1.5	1	1
104E137	1.5	2.5	2
104E9	3.5	2.5	3
108E9	3.5	4	4
40E8	5	5	5

[a] Data of Table 7.3, reworked; 1 = least aggressive, 5 = most aggressive.

TABLE 7.6

Reversed Reaction Types Characteristic of Vertical Resistance When Two Host Genotypes Are Inoculated with Two Races of the Pathogen

Host	Race 1	Race 2
Host with gene R1	High reaction type[a]	Low reaction type
Host with gene R2	Low reaction type[b]	High reaction type

[a] For example, reaction type 4 on the conventional scale for cereal rusts, indicating susceptibility.

[b] For example, reaction type 0; indicating resistance.

race 1 but not race 2. Ranked in order of decreasing resistance, the R2 genotype would be ranked on top, and the R1 genotype at the bottom when tested with race 1; when tested with race 2 the R1 genotype would be ranked on top, and the R2 genotype at the bottom. Race 1 is virulent for gene R1, but avirulent for gene R2; race 2 is avirulent for gene R1, but virulent for gene R2. The order of virulence is reversed from the R1 genotype to the R2 genotype.

Biometric methods exist for measuring the degree of uniformity between two or more rankings, with tests of significance based on ordinal numbers 1, 2, 3, . . ., which denote order or rank. They lose information about how close the ranked members are within the scale. In this respect they are inferior to the familiar statistical methods that use cardinal numbers. But ranking methods have the advantage that ranking is independent of the stretching of the scale of measurement, and the methods avoid the requirements that encumber the analysis of variance. More use might be made of them when disease is measured as a percentage or on an arbitrary scale as 1 = little disease and 10 = much disease.

7.5 VERTICAL RESISTANCE EFFECTIVE ONLY
AGAINST INITIAL INOCULUM

Vertical resistance (with exceptions to be noted) is effective only against the initial inoculum reaching a field or plot. Paraphrased, vertical resistance is effective only during the first cycle of disease within a field or plot. Suppose that the cultivar has the resistance gene R1 and that inoculum in the form of airborne spores belongs to various races: races (0), (1), (2), ... Gene R1 is effective against races (0), (2), . . ., and all other races without 1 in their designation. It is ineffective against race (1) and all other races with 1 in their designation. Therefore, in the first cycle of disease races (0), (2), . . ., are screened out, and only races (1) and other races with 1 in their designation remain. The second cycle of disease within the field is therefore caused solely by race (1) and other races with 1 in their designation. To them the cultivar has no vertical resistance. Vertical resistance has fallen away, and only horizontal resistance remains in the second cycle onwards. This rule is subject to two exceptions discussed in the last paragraph of this section.

The same argument applies when the pathogen moves from field to field or plot to plot during the course of an epidemic. Vertical resistance can be effective only against the initial inoculum, and not against inoculum originating from within the fields or plots themselves. Put differently, genes give vertical resistance only against inoculum coming from host plants with different genes (Vanderplank, 1963, pp. 197–198).

Ahn and Ou (1982a,b) studied the resistance of rice cultivars to blast disease caused by *Pyricularia oryzae*. Rice cultivars differ greatly in the number of races of *P. oryzae* to which they are resistant, and there are fewer blast lesions on cultivars resistant to many races. Ahn and Ou determined how the proportion of races to which a cultivar is resistant is correlated with the increase of disease during overlapping secondary and tertiary cycles of infection. Continuous plantings with overlapping growth periods were made with seven rice cultivars. There were three plantings at 15-day intervals. Each plot consisted of five 1-m rows, 10 cm between rows, 20 cm between plots, and with 2 m between cultivars. The first planting was inoculated artificially. Disease was recorded as the number of blast lesions per 100 cm² of leaf area. Disease ratios recorded in Table 7.7 were calculated without correction for multiple infections; correction would not change the main conclusions.

First, resistance in the cultivars tested by Ahn and Ou was mostly vertical. The cultivars differed widely during the first cycle of disease, the differences reflecting total resistance both vertical and horizontal. But the differences in horizontal resistance, as shown by the entries in the last two columns of Table 7.7, were small and inconsistent, from which one concludes that

TABLE 7.7

Development of Blast Disease Caused by *Pyricularia oryzae* in Seven Rice Cultivars in Three Continuous Plantings with Overlapping Growth Periods

Cultivar	Percent isolates resisted[a]	Lesions per 100 cm² of leaf area[b]			Disease ratios[e]	
		First cycle[c]	Second cycle[d]	Third cycle[d]	Second : first cycle	Third : second cycle
Telep		13	27	62	2.1	2.3
NP-125	79	14	28	106	2.0	3.8
CI 5309	76	35	80	222	2.3	2.8
Dular	71	22	53	155	2.4	3.7
Ramidad str. 3	36	53	153	279	2.5	2.1
Peta	24	129	350	571	2.7	1.6
Khao-tah-haeng 17	5	323	678		2.1	

[a] Percentage of isolates out of 103 to which the cultivars were resistant when they were artificially inoculated. Calculated from the data of Ahn and Ou (1982a).

[b] Determined from the graphical data of Ahn and Ou (1982b).

[c] The amount of disease developed after artificial inoculation. This reflects both vertical resistance and horizontal resistance.

[d] Disease at 15-day intervals after the first cycle.

[e] The ratios reflect horizontal resistance only, on Ahn and Ou's evidence that each cultivar was resistant or susceptible to each particular race to which the isolate belonged.

vertical resistance was the main component of the total resistance. This conclusion agrees well with the evidence about the percentage of isolates resisted by the various cultivars. Second, there is no suggestion in the evidence of any relation between a cultivar's horizontal and vertical resistance. Horizontal and vertical resistance vary independently.

The rule underlying this section, that vertical resistance is effective only during the first cycle of disease in a field, has two exceptions. The rule does not apply to fields of multilines or mixed varieties, i.e., the host plants must be genetically uniform in vertical resistance. The rule does not apply when the vertical resistance is only partially effective against avirulent races of the pathogen (see Section 7.7).

7.6 INTERRUPTED OR UNINTERRUPTED PRESENCE OF DISEASE OR INOCULUM

Vertical resistance genes do not protect a field against spores or other inoculum generated within the same field. This is the principle underlying the previous section. It follows that vertical resistance genes are effective only when the supply of inoculum or disease is interrupted. Again, we exclude vertical resistance that is only partially effective against avirulent races of the pathogen. Consider some examples.

7.6.1 *Puccinia graminis tritici* in North American Spring Wheat

Now that the alternative host *Berberis* has been effectively eradicated, *Puccinia graminis tritici* does not survive the winter in the spring wheat area of North America. There is a yearly interruption, and every summer new inoculum arrives from the south, having overwintered on different genotypes from those in the north. Spores of various races arrive, but the resistant spring wheats eliminate them in the first cycle of disease. Protection is efficient and for more than 30 years has proved to be lasting. Here we see vertical resistance at its best.

The evidence of success is strong. Green and Campbell (1979) list 10 wheat cultivars licensed in Canada during the last 30 years and describe their performance in resisting stem rust, their resistance being effectively vertical. Only Canthatch, licensed in 1959, and Pitic 62 licensed in 1969, have become susceptible. Of the rest, 7 became cultivated over areas exceeding a million hectares a year in their heyday without becoming susceptible to stem rust. They are Selkirk licensed in 1953, Pembina licensed in 1959, Manitou licensed in 1965, Neepawa licensed in 1969, Napayo and Glenlea licensed in 1972, and Sinton licensed in 1975.

The very success of these cultivars has had unfortunate repercussions on plant breeding for resistance to disease in that it has been wrongly inferred that what has been done with vertical resistance to wheat stem rust in Canada and the United States ought to be a model to be followed elsewhere. It has not been realized that in terms of the interrupted presence of disease or inoculum, stem rust of North American spring wheat is unusual. The North American wheat stem rust model is exceptional.

7.6.2 *Erysiphe graminis hordei* in Western Europe

Erysiphe graminis hordei, the cause of powdery mildew in barley, has effective means of uninterrupted local perpetuation. In hot summers it can oversummer as cleistothecia (sexual fruiting bodies). In the relatively mild maritime climate of western Europe it overwinters asexually on fall-sown crops and volunteer plants.

Uninterrupted local survival, implying as it does that fields will be infected by inoculum coming from infected plants of the fields' own resistance genotypes, is inimical to the use of vertical resistance. The resistance given by the vertical *Ml* genes has proved to be unstable and associated with the usual boom-and-bust cycles. In Britain, gene *Ml-g* was widely used in barley cultivars that became popular in the 1960s; corresponding virulence became prevalent in *E. graminis hordei*. The addition of gene *Ml-a6* gave relief that proved to be temporary; in 1964 surveys revealed no isolates of *E. graminis hordei* virulent for genes *Ml-g* and *Ml-a6* in combination; by 1968, 84% of the isolates were virulent for these genes in combination (Howard *et al.*, 1970).

Fungicides have come to the rescue of barley growers. Nevertheless, barley breeders have not entirely abandoned vertical resistance. There is hope that stabilizing selection will act against virulence for appropriate resistance genes combined within a cultivar (Wolfe and Barrett, 1977). The replacement of single barley cultivars by variety mixtures (Wolfe, 1978) exploits vertical resistance genes in another way when there is uninterrupted existence of disease and inoculum; here too stabilizing selection against "super races" of the pathogen is necessarily involved.

7.6.3 *Phytophthora infestans* in Potatoes

Phytophthora infestans, causing blight in potatoes, soon establishes an uninterrupted local existence by infecting tubers. A race of *P. infestans* can coexist with a compatible genotype of potato without the movement from genotype to genotype that vertical resistance needs. Ths history of blight in the potato cultivar Kennebec is typical. Kennebec has the gene *R1*. It was released in the United States in 1948. Although it was not the first variety with the gene *R1* to be released in North America, it was the first to become widely grown. At the time of its release virulence for the gene *R1* was so rare that Kennebec was highly resistant; fields of Kennebec either escaped blight altogether or developed a few lesions late in the season, too few and too late to do much harm. Farmers had no need to use fungicides. In this boom period production of Kennebec increased fast. By 1954, Kennebec and other *R1* cultivars had increased and now accounted for 6.3% of the certified seed grown in Maine. With this increase came trouble and the start of the bust period of the resistance. In 1954, slight to moderate infection was found in widely separated fields of Kennebec as early as the beginning of July (Webb and Bonde, 1956). With the advance of summer a full epidemic of blight developed on Kennebec. At the Aroostook Farm, Presque Isle, Maine, blight was so severe that it caused 90% defoliation in unsprayed plots before September 4 (Stevenson *et al.*, 1955). Kennebec was no longer resistant, and races able to attack it were evidently abundant. This suggested that Kennebec was already grown widely enough for appropriate inoculum to overwinter in substantial amounts. This suggestion was confirmed by Webb and Bonde (1956). They tested 56 isolates from 15 cull piles of diseased tubers thrown out near potato fields in the spring of 1955. Of the 56 isolates 30 could attack Kennebec. The reason for the downfall of Kennebec's resistance was obvious: Kennebec's increasing popularity among farmers had brought with it an increase of Kennebec-attacking races, and through infected cull piles and seed potatoes the appropriately virulent races coexisted uninterruptedly with Kennebec itself.

The downfall of vertical resistance has been so complete that potato farmers rely on fungicides to control blight, and potato breeders insofar as they are concerned with resistance to blight have turned to horizontal resistance.

7.6.4   *Hemileia vastatrix* in Coffee

Hemileia vastatrix, causing rust in coffee, soon establishes an uninterrupted local existence, perpetuating itself the year round on coffee trees. The pathogen is a biotroph and inoculum is reduced when infected leaves fall, but the reduction is never enough to break the continuity of disease. Vertical resistance genes can be effective only when *H. vastatrix* first arrives. Against any genotypes of the pathogen that survives the first generation of attack, the vertical resistance genes are necessarily ineffective; subsequent generations can be resisted only horizontally.

When *H. vastatrix* was first found in Brazil in 1970, the race was identified as race II (Schieber, 1975) which is virulent for the resistance gene *SH5*.

By 1972 a new race, race XV, was identified. It is virulent for genes *SH4* and *SH5*. Other races quickly appeared thereafter. New coffee plantations need 10 years to reach high productivity, so the chances that coffee breeders can keep vertical resistance ahead of the fungus seem slim.

Coffee rust illustrates the difficulty of using vertical resistance against diseases of trees; it is practically impossible to meet the requirement of vertical resistance that the pathogen be forced to move from one host genotype to another.

7.6.5 *Synchytrium endobioticum* in Potato

Synchytrium endobioticum, the cause of wart disease of potatoes, illustrates how slow multiplication of disease can save vertical resistance. The resting sporangia of *S. endobioticum* remain viable in soil for many years, and inoculum can exist uninterruptedly in a field under normal crop rotation. Nevertheless, vertical resistance has been very successfully used.

Synchytrium endobioticum is favored by cool, moist conditions. It is endemic in the Andes, the ancestral home of *Solanum tuberosum*. Genes for vertical resistance are common in the species. *Synchytrium endobioticum* is spread mainly by infected seed potatoes but also by soil, compost, or manure. In most countries where it occurs legislation enforces the planting of resistant potato cultivars, and this has proved to be successful. Races of the fungus occur, and new races have caused the replacement of potato cultivars in Germany, Eastern Europe, and Newfoundland; but no great difficulty has been found in the replacement.

There are at least three reasons why wart disease has been so successfully controlled by vertical resistance. First, resistance occurred within *S. tuberosum* itself and was readily available. It occurred even when it was not selected for. For example, the cultivar Sebago was used in Newfoundland as a resistant variety. Yet Sebago was bred in the United States where wart disease does not occur, and there is no evidence that Sebago was ever tested for resistance against wart disease before it was released as a commercial cultivar. Certainly, resistance to wart disease has played no part in the worldwide success of this cultivar, except in Newfoundland. So, too, in Europe when wart disease was first found many agronomically successful cultivars already in cultivation or in the pipeline were resistant. Second, because resistant cultivars were available, there were few difficulties in enforcing the planting of resistant cultivars by legislation. The legislation brought about a rapidly dwindling supply of inoculum and therefore a rapidly dwindling chance of mutation from avirulence to virulence. Third, the fungus spreads slowly and lacks means of dispersal through the air. There was plenty of time to contain mutation to virulence.

Compare the success of vertical resistance against *Solanum endobioticum* with the failure of vertical resistance against *Phytophthora infestans* in the potato. Vertical resistance genes for *S. endobioticum* were already in the cultivated potato, in agronomically successful varieties. Vertical resistance genes for *P. infestans* had to be sought elsewhere, especially in the hexaploid weed *Solanum demissum*. Available cultivars resistant to wart disease allowed legislation to be passed to prohibit the planting of susceptible cultivars, thereby reducing inoculum and inoculum-based mutation to virulence. Against *P. infestans* corresponding legislation would have been totally impracticable. Third, *Synchytrium endobioticum* spreads slowly, *P. infestans* swiftly. The swift spread of *P. infestans* is illustrated by the spread of races able to attack the cultivar Pentland Dell which has genes *R1*, *R2*, and *R3*. Malcolmson (1969) has given details. Until 1967 Pentland Dell was for all practical purposes a resistant variety. In 1967 blight appeared in Pentland Dell in Cornwall at the end of July. Extensive infection was soon noted in Cornwall and Devon, and within a month reports of crops with 75% of the foliage destroyed were common. By the end of 1967 blight in Pentland Dell had appeared in 14 counties of England and Wales. In 1968 it appeared even as far away as Scotland. It is immaterial whether this spread was all from a single mutation or whether mutants arose in several places. Either way, potato breeders were faced with an impossible task of containing *P. infestans*.

7.6.6 Discussion

Five examples of the use of vertical resistance have just been given. Against two diseases, wheat stem rust in North America and wart disease of potatoes, vertical resistance succeeded, but for very different reasons. Against two of them, potato blight and coffee rust, vertical resistance has so far failed. Against powdery mildew of barley, success has been incomplete. These results have an ecological background and could have been predicted from a knowledge of the diseases. Certainly, there are instances in which the use of vertical resistance can be seen in advance to be hopeless, as against disease in perennial, genetically uniform crops. It is, however, not implied that success or failure in using vertical resistance depends solely on ecological background. The abundance, availability, and quality of vertical resistance genes are other factors.

The many failures and boom-and-bust cycles have given vertical resistance an undeservedly bad name. Vertical resistance can be very successful but must be used selectively. The essential rule, it would seem, is to assume in advance that mutations from avirulence to virulence will occur, and then to accept or reject the use of vertical resistance genes on the ecological background.

7.7 VERTICAL PARTIAL RESISTANCE

In previous sections vertical resistance has been taken as complete. Avirulent races of the pathogen produce no symptoms or small, hypersensitive flecks or small lesions that remain sterile. But vertical resistance may only be partial. Avirulent races produce lesions that may be fertile, but less fertile than the lesions produced by virulent races. Vertical resistance that is partial has often been confused with horizontal resistance, which is commonly partial. Partial resistance is *not* an acceptable synonym for horizontal resistance, although it is regrettably often so used.

A common classification of reaction types was given by Stakman *et al.* (1962) for *Puccinia graminis tritici* and is reproduced in Table 7.8. Similar classifications have been used for other fungi: Johnston and Browder's classification for *P. recondita* drops class 2 and stresses class X. Intermediate reactions, indicating partial resistance, are common. For example, in the reactions listed by Stakman *et al.* (1962) for 294 races of *P. graminis tritici* on the wheat cultivar Marquis more than one-third are of types 2 or X, with or without + or − signs.

Vertical partial resistance can be accumulated. Samborski and Dyck (1982) studied the accumulation of vertical partial resistance in wheat to *P. recondita.* A number of genes in wheat that singly give intermediate levels

TABLE 7.8

Conventional Infection Types Produced by *Puccinia graminis tritici* on Wheat Seedlings and Their Interpretation[a]

Infection type[b]	Seedling reaction
0	Immune. No visible indication of infection
0;	Nearly immune. Hypersensitive flecks present
1	Very resistant. Uredia minute, surrounded by necrotic areas
2	Moderately resistant. Uredia small to medium, usually in green islands with chlorotic or necrotic borders
X	Heterogeneous, intermediate resistance. Uredia variable, sometimes including all infection types
3	Moderately susceptible. Uredia medium in size, sometimes with chlorotic areas but no necrosis
4	Very susceptible. Uredia large and often coalescing, sometimes with chlorosis but no necrosis

[a] From data of Stakman *et al.* (1962).

[b] Variation within an infection type is indicated by +, −, c (much chlorosis), and n (much necrosis) signs. Thus, 2 + + might overlap 3 = as an indicator of resistance or susceptibility.

of resistance were studied in various combinations. Highly resistant selections of wheat were obtained from an intercross of four single-gene lines, each giving a low level of resistance. Increased seedling resistance was noted in backcross lines of Neepawa wheat with resistance genes *Lr11* and *Lr30* when compared with the reaction of either gene singly. Races avirulent for both genes showed reduced infection types (i.e., higher resistance) when compared with races avirulent for one of the genes but not the other.

Breeding programs have been launched to accumulate partial resistance on the unfounded assumption that it is necessarily horizontal. Unless horizontal and vertical partial resistance are distinguished (see Section 9.8), the programs may end with high vertical resistance that is just as vulnerable to boom-and-bust cycles as the more usual form of vertical resistance that uses pseudomonogenes. Samborski and Dyck's (1982) results warn about this. Races of *P. recondita* virulent for genes *Lr11* and *Lr30* left the Neepawa backcross line with *Lr11* and *Lr30* just as susceptible as the backcross lines with either *Lr11* or *Lr30* singly or Neepawa itself. There was no resistance remnant (see Chapter 8).

The term "cumulative resistance" is used in this section instead of the plainer term "additive resistance" to avoid confusion. Additive resistance has a special meaning, signifying that in resistance the heterozygous hybrid is intermediate between the homozygous resistant parent and the homozygous susceptible parent (see Chapter 3). This meaning is not relevant to this section.

7.8   ADULT-PLANT RESISTANCE: SCHEIBE'S RULE

On present evidence there seems to be little point in using genes for partial vertical resistance if genes for complete vertical resistance are available. One exception might be noted. Partial vertical resistance in young seedlings often goes with substantially complete vertical resistance in the adult plant, in which case it can be useful as field resistance.

Scheibe (1930), in Germany, stated a rule that if a wheat variety is either very resistant or very susceptible to *P. recondita*, there is no change of reaction as the plant ages; the change from seedling susceptibility to adult-plant resistance is confined to varieties that are only moderately susceptible or have an intermediate reaction type. In crosses between susceptible and resistant wheat varieties, the homozygous F_2 progeny are either resistant or susceptible throughout life, whereas the heterozygous F_2 progenies may be susceptible as seedlings but resistant later in life.

It is an implication of Scheibe's rule that genes for vertical adult-plant resistance are just ordinary vertical resistance genes with effects conditioned by ontogenic change.

Results from North America are at variance with those of Scheibe in that they tend to show a greater increase in resistance from seedlings to adult plants. Table 7.9 is based on wheat leaf rust data obtained by Dyck *et al.* (1966) from the F_2 plants of a Thatcher × Exchange cross. The wheat cultivar Exchange has adult-plant resistance to *P. recondita* race 5; cultivar Thatcher has not. The table shows that plants resistant, with reaction types $1+$ to 2 as seedlings, are mostly also resistant, with reaction types from 0; to 2 as adult plants. Only a few (seven) were intermediate in resistance at the later stage. (These few could have been eliminated if reaction types $2+$ and $2++$ were classified as indicating resistance, as Dyck *et al.* classified them in calculating segregation ratios.) None was susceptible, with reaction type 4, at the later stage. All this agrees with Scheibe's rule. Plants with intermediate resistance, with reaction types $2++$ to 3 in the seedling stage, were also mostly resistant as adults. None was susceptible, with reaction class 4. The main departure from Scheibe's rule was in the change in plants classified as susceptible as seedlings. Many, contrary to Scheibe's rule, were resistant as adults; most had an intermediate resistance.

The part of Scheibe's rule that states that heterozygous F_2 progenies may be susceptible as seedlings but resistant later is better followed in the data of Dyck *et al.* (1966) from backcrosses of Exchange to Thatcher. In the F_3 all adult plants with seedling reaction types from 0; to 2 were homozygously

TABLE 7.9

Wheat Leaf Rust: The Relation between Seedling and Adult-Plant Resistance in Plants of an F_2 Population from a Thatcher × Exchange Cross[a]

	Number of plants with the indicated adult-plant reaction		
Seedling reaction	0; to 2[b] (resistant)	2+ and 2++ (intermediate)	4[c] (susceptible)
1+ to 2	131	7	0
2++ to 3	151	16	0
4	67	86	47

[a] From data of Dyck *et al.* (1966), using race 5 of *Puccinia recondita*.

[b] Reaction types 0;, 0;1−, 0;1, and 2 in the data of Dyck *et al.* were merged to indicate resistance. Thus, there were 131 plants that had a seedling reaction type from 1+ to 2 and an adult-plant reaction type from 0; to 2.

[c] Dyck *et al.* did not record any plants with reaction type 3 in the adult-plant stage.

resistant; those with a 2+ seedling reaction type segregated into three groups: resistant, heterozygous (type 2+ +), and susceptible; whereas those with a 4 reaction type were homozygously susceptible. It is also followed in the data of Athwal and Watson (1957). They found two genes in the wheat cultivar Mentana conditioning resistance to *P. recondita*. Both genes condition resistance at both the seedling stage and in the field. Both genes are recessive in the seedling stage, whereas one showed reversal to dominance later in the field.

The evidence taken together shows a tendency for leaf rust vertical resistance genes to be more effective in adult plants than in young seedlings. The North American evidence for both leaf rust and stem rust suggests that this tendency is more marked for some vertical resistance genes than others. These are the genes singled out as genes for adult-plant resistance. For example, genes *Lr13* and *Sr2* are genes for vertical adult-plant resistance.

The interaction between genes for adult-plant resistance and plant age does not reduce the interaction between host and pathogen. The resistance remains vertical and is "lost" in the presence of virulent populations of the pathogen. On present evidence there is no advantage in using vertical adult-plant resistance, i.e., resistance absent or weak in young plants, instead of vertical resistance that operates throughout the life of the plant unless there is a shortage of available genes.

Opposite physiological trends confuse the study of adult-plant resistance. As wheat plants grow old, they tend to lose sugar concentration in leaves, stems, and chaff, a process known to straw-chewers down the ages and investigated chemically (Anonymous, 1937). To cereal rust diseases, decreased sugar means increased resistance, and this might well be a process underlying Scheibe's rule. In the opposite direction, many cereal crops grow from winter to summer, and increasing temperature tends to increase reaction types, expecially those that are intermediate. As a result, adult-plant resistance in studies at constant temperature is likely to be clearer than when observed in the field.

7.9   HIGHER-ORDER INTERACTIONS

The past three sections of this chapter have dealt with vertical resistance. The rest of this chapter deals mostly with horizontal resistance.

Horizontal resistance is defined in a system of only two variables, host and pathogen. By definition of horizontal resistance there is no interaction between them. But this does not preclude the existence of higher-order interactions, i.e., of interaction with other variables, all compatible with horizontal resistance.

Consider an example from potato blight. One of the manifestations of horizontal resistance to *Phytophthora infestans* is a lengthened period needed for a spore to germinate and the sporeling to establish infection. During the whole of this period free water is necessary, and the length of the period depends on temperature. This form of resistance is specially effective when wet periods are relatively short; it loses much of its advantage in weather with long spells of rain and mist. Thus, even though there may be no first-order interaction, host × pathogen, there can be second-order interactions, host × pathogen × wet periods or host × pathogen × temperature, and a third-order interaction, host × pathogen × wet period × temperature. For example, the order of rank of potato cultivars for resistance to blight could differ with the wetness of the climate. Still higher orders are conceivable.

Robinson (1979) and Kulkarni and Chopra (1982) have discussed the environment (including interplot interference) as a cause of higher-order interactions involving host and pathogen.

Higher-order interactions occur even without environmental variation when resistance is measured as a lower infection rate. Polycyclic disease progresses cyclically in waves, i.e., with a cyclically varying infection rate. Examples were given by Vanderplank (1975, pp. 95–96). With resistance measured by infection rates, a second-order interaction, host × pathogen × level of disease, can be expected even in the absence of a host × pathogen interaction, because resistance affects the level of disease. The misinterpretation of an interaction, host × pathogen × level of disease, as a host × pathogen interaction occurs in the analysis by Latin *et al.* (1981) of the infection rate of *P. infestans* in four potato cultivars.

7.10 HOW REALISTIC IS THE DEFINITION OF HORIZONTAL RESISTANCE?

In the face of possible higher-order interactions, how realistic is a definition based on the absence of a first-order interaction? Two cases arise, first, when there are substantially only two relevant variables (the topic of this section) and second, when resistance automatically introduces other variables (the topic of the next section).

Suppose maize breeders release a new maize hybrid with more resistance to stalk rots. Maize farmers will plant it, and it may supersede other hybrids that are agronomically similar. But it is unlikely to bring about large changes in the pattern of maize farming. The cultivated area under maize, crop rotations, and general agronomic practices will stay much the same; they are dictated by economics and farm practices. There will be variation of host plants by supersession of hybrids, but no substantial change of other variables. It is against a background substantially invariable for other factors that one must gauge the historic stability of resistance to maize to stalk rots and for that matter most other diseases.

What holds for maize holds for the other major field crops. The pattern of farming is not easily ruffled by a new cultivar even if it becomes established, and definitions of resistance within a two-variable system are realistic enough.

7.11 A THIRD VARIABLE

In contrast with what has just been said, there are instances of the introduction of resistant cultivars changing the pattern of farming. With resistance available, farmers respond by saving on chemicals, shortening crop rotations, or starting cultivation in areas previously thought to be dangerous. Tomato farming in warm regions where fusarium wilt is endemic is an example of how rotations were shortened. Fusarium wilt is caused by *Fusarium oxysporum lycopersici*. Good resistance to it became available in tomato cultivars in the early 1940s. Before that date, for fear of fusarium wilt, farmers in warm areas used long crop rotations or sought land not previously used for growing tomatoes. After that date, they felt safe in using short rotations. They introduced a third variable, with a potential second-order interaction, host × pathogen × crop rotation. There have of course been other changes since the 1940s, such as the widespread use of chemicals for soil fumigation, but the host × pathogen × crop rotation interaction can usefully be singled out for discussion.

The common race of *F. oxysporum lycopersici* was, and in most areas still is, race 1. Gene *I* from *Lycopersicon pimpinellifolium* conditions resistance to this race. Gene *I* was available in the tomato cultivar Pan America, released in 1941, and is now available in many others. Race 2 pathogenic for gene *I*, was first found in Ohio in 1945 (Alexander and Tucker, 1945) as an isolate from a wilting tomato plant without gene *I* at a time when most tomato plants still lacked this gene. This suggests that race 2 is either a widespread but infrequent constituent of the fungus population or that race 1 mutates easily to race 2. Later, Gerdemann and Finley (1951) inoculated a field that had never grown tomatoes before. They inoculated it for 5 successive years with three single-spore isolates of race 1. Another field was inoculated for 2 years. Tomatoes of different varieties were planted in the fields. Varieties without the gene *I* wilted. A few plants with this gene also wilted. From 12 of them the pathogen was isolated and proved to be of race 2. The race had evidently arisen from race 1 by mutation. There was a surprisingly long lull, from 1945 to 1960, before race 2 was reported in commercial fields. In 1960

Stall (1961) found it in Florida. Since then it has been widely reported in the United States, the Middle East, North Africa, Latin America, the Balkans, and southern Europe.

Although sporadic occurrences of race 2, causing no substantial harm to farmers but consternation among plant pathologists, have been recorded where tomatoes had previously not been grown, the threat by race 2 has been closely related to growing tomatoes in the same soil in quick succession. Crop rotations have become very short compared with those in the days before gene *I* was available. Cirulli and Alexander (1966) note that three of the four early isolates of race 2 in Florida were from plants grown in soil that had had at least 2 consecutive tomato crops. Walter (1967) notes that varieties with gene *I* enable 10 successive crops of tomatoes to be grown on the sandy soils of Florida where 2 crops would have been the limit using varieties without this gene. Crill *et al.* (1972) report that infection by race 2 reached 43–74% when tomatoes with gene *I* were grown 2 years in succession, as compared with 0–3% when grown only once during a comparable period. In Israel Katan and Ausher (1974) found that the build-up of race 2 in fields was related to the frequent cropping of tomatoes. The reality of the interaction, host × pathogen × crop rotation, is not in doubt. Historically, race 2 is a mutant or constituent of the fungus population that was found sporadically in insubstantial amounts while tomatoes were grown in long rotations or in soil not previously planted to tomatoes; the race needed short crop rotations, with tomatoes in frequent succession, before it became a major threat.

7.12   QUALITATIVE AND QUANTITATIVE VARIATION IN HOST AND PATHOGEN

Qualitative variation in host and pathogen is implied in vertical resistance. Quantitative variation is implied in horizontal resistance and fits the evidence about races 1, 2, and 3 of *F. oxysporum lycopersici* and the resistance genes of tomato.

Gene *I-2* (Stall and Walter, 1965), also from *L. pimpinellifolium*, conditions resistance to race 2 of *F. oxysporum lycopersici*. Tomato cultivars with gene *I-2* are available and have been widely used, although cultivars with gene *I* are still popular even where race 2 is known to occur.

Another race, race 3, of *F. oxysporum lycopersici* is known (Alexander and Hoover, 1955; Tokeshi *et al.*, 1966). It is pathogenic for *both* genes *I* and *I-2*. It appears to be rare and has not been reported as being destructive. Correspondingly, another gene exists in *L. pimpinellifolium* and tomato line

PI 126915-1-8-1. Selfed plants of this line are homozygously resistant to *both* races 1 and 2 (Cirulli and Alexander, 1966). The evidence is thus for quantitative rather than qualitative pathogenicity in the fungus and resistance in tomato.

Table 7.10 compares qualitative and quantitative reactions. At least two resistance genes, *R1* and *R2*, and two pathogenicity genes, *p1* and *p2*, are needed if qualitative and quantitative variation is to be distinguished. Two resistance genes give 1 degree of freedom for resistance; two pathogenicity genes give 1 degree of freedom for pathogenicity; and two resistance genes and two pathogenicity genes give 1 degree of freedom for host × pathogen interaction. In the known data for fusarium wilt of tomato there is a single degree of freedom for interaction. Race 3 (i.e., *p2*) is pathogenic for both gene *R1* (i.e., gene *I*) and gene *R2* (i.e., gene *I-2*). This indicates quantitative variation. Had variation been qualitative, race 3, with *p2*, would have been pathogenic for gene *R2* but not *R1*, the reverse of race 2, with *p1*, which is pathogenic for *R1* but not *R2*. Until such time as a race is found that attacks

TABLE 7.10

Reaction of Plants When Three Resistance Genes *R1*, *R2*, and *R3*
and Three Pathogenicity Genes *p1*, *p2*, and *p3* Are Arranged
Qualitatively (A) and Quantitatively (B)

	Resistance genes		
Pathogen	*R1*	*R2*	*R3*
A. Qualitative arrangement			
Avirulent	Resistant	Resistant	Resistant
p1	Susceptible	Resistant	Resistant
p2	Resistant	Susceptible	Resistant
p3	Resistant	Resistant	Susceptible
B. Quantitative arrangement			
Avirulent	(Resistant)[a]	(Resistant)	(Resistant)
p1	(Susceptible)	(Resistant)	(Resistant)
p2	(Susceptible)	(Susceptible)	(Resistant)
p3	Susceptible	Susceptible	Susceptible

[a] The reactions in parentheses are known for *Fusarium oxysporum lycopersici* in tomatoes. The avirulent race is race 1; *p1* stands for race 2; and *p2* for race 3 of this fungus. The conventional numbering of these races, with race 1 (instead of race 0) standing for the avirulent race, is confusing. Genes *R1*, *R2*, and *R3* stand for genes *1*, *1-2*, and the gene in PI 126915-1-8-1, respectively.

plants with gene *I-2* but not those with gene *I*, the resistance given genes *I* and *I-2* must be considered horizontal.

Note that in Table 7.10 the avirulent race (race 1 of *F. oxysporum lycopersici*) does not contribute to the degrees of freedom.

7.13 UNCLASSIFIED RESISTANCE

Monogenic resistance in cabbage to cabbage yellows disease caused by *F. oxysporum conglutinans* has remained stable. Without a race pathogenic for the resistance gene, there is no degree of freedom to test for interaction. One may suspect that resistance in cabbage to *F. oxysporum conglutinans*, like that in tomato to *F. oxysporum lycopersici*, is probably horizontal, but there is no experimental evidence for this. Without relevant variation in the pathogen, definitions of horizontal and vertical resistance automatically fall away.

Stable monogenic or oligogenic resistance is common. It is found widely against fungal, bacterial, and viral diseases. Examples are milo disease of sorghum caused by *Periconia circinata*, bacterial wilt of tobacco caused by *Pseudomonas solanacearum*, tobacco mosaic caused by tobacco mosaic virus, and raspberry ringspot caused by tomato black ring virus. In the absence of appropriate variation to give degrees of freedom for interaction, the resistance remains unclassified. Meanwhile, the general advice to plant breeders is simple: When monogenic resistance is available, use it; it has a good chance of being stable.

Degrees of freedom are not peculiar to any particular method of biometric analysis. Their number is the number of independent comparisons that can be made in any set of data. How one makes the comparisons is not an issue. The term degree of freedom has long been used in science quite apart from biometry; for example, it enters the Phase Rule in physical chemistry, and the rule has been known for more than a century.

7.14 PSEUDOSPECIFITY

True specificity is qualitative. On the evidence of Section 7.12 races 1 and 2 of *F. oxysporum lycopersici* vary quantitatively. One should therefore refer to them as pseudospecific. Pseudospecificity is introduced into horizontal resistance by a third (or higher) variable.

Consider in tomato fusarium wilt the second-order interaction, host × pathogen × crop rotation. There is a very strong first-order interaction, host × crop rotation. The introduction of host plants with gene *I* allowed crop rotations to be shortened; successive crops of tomato were grown, a practice too dangerous to be followed without gene I. There is a strong first-order interaction, pathogen × crop rotation. Shorter rotations brought in race 2 as a major race. But these interactions are irrelevant to the definition of horizontal resistance, which is the absence of an interaction, host × pathogen, all other variables being equal.

7.15　REMNANTS OF HORIZONTAL RESISTANCE

Horizontal, unlike vertical, resistance keeps by definition a remnant even in the presence of more pathogenic races. To continue to use crop rotation to illustrate a third variable, if variety A is more resistant than variety B in long rotations, it will be more resistant in short rotations, even if to maintain the order of rank, variety A better than B, one must translate "more resistant" as "less susceptible." To cite fusarium wilt of tomato, plants with gene I are less susceptible than plants without this gene, even in the presence of race 2 which is pathogenic for gene I. Gene I leaves a remnant. Farmers continue to use gene I widely, despite the presence of race 2.

In Israel, Katan and Wahl (1969) studied the effect of the presence of gene I in tomatoes inoculated with race 2 of *F. oxysporum lycopersici*. Seedlings of the cultivar Marmande, which lacks gene I, collapsed within a few days of inoculation; seedlings of the cultivar Homestead F-M, which has gene I, became infected but were more tolerant, showing stunting rather than collapse.

Remnants must be sought in the appropriate form of resistance. In tomato, resistance given by gene I does not prevent infection but blocks the fungus in the vascular tissue after infection. Remnants of resistance must therefore be sought in processes after infection, which is the case.

7.16　HORIZONTAL RESISTANCE AND STABILIZING SELECTION

Horizontal resistance, unlike vertical resistance, is stable in a two-variable, host–pathogen system without the need for stabilizing selection (other than that which operates in all populations in all biological systems). Horizontal resistance is commonly introduced into field crops like maize without significantly affecting other variables (see Section 7.10); consequently, stabilizing selection does not greatly enter discussions of horizontal resistance.

It is when the introduction of resistant varieties disturbs other variables that stabilizing selection has a major role in maintaining stable horizontal resistance. Stabilizing selection then operates against increased aggressiveness in the pathogen when this increased aggressiveness is coupled, as one expects it to be, with decreased fitness of the pathogen to survive outside the host plants. Fitness is defined (as Darwin defined it) as success in leaving progeny; the fitness of a race of a pathogen is measured by the race's maintaining itself as a major constituent of the pathogen's population. Fusarium wilt of tomatoes is a clear example. Historically, race 1 was, and commonly still is, the most abundant race, i.e., the fittest race. Race 2 with aggressiveness for one resistance gene is rarer. Race 3 with aggressiveness for two resistance genes is rare to the point of obscurity. The lack of fitness of race 2 relative to race 1 to survive outside the plant stabilizes the resistance given by gene I except in short rotations, when the fitness of the pathogen to survive within the plant, i.e., parasitically, becomes more important. Probably, the lack of fitness of race 3 relative to races 1 and 2 has stabilized the resistance given by gene I-2, but about this we can only guess at present.

8

Remnants of
Resistance

8.1 THREE QUESTIONS

There is a common experience, usually called the boom-and-bust cycle. The plant breeder releases a new vertically resistant cultivar to farmers. Initially, resistance is maintained and the cultivar is successful. This is the period of boom in the cultivar's career. Later, the pathogen changes from avirulence to virulence for the vertical resistance genes. Resistance is "lost." This is the period of bust.

Is the resistance all lost? Is there a remnant? Despite the bust, is the farmer better off than he was before, or worse? There are really three distinct questions.

First, what happens when the vertical resistance has a heterogeneous background of horizontal resistance? When the vertical resistance is lost, there is a remnant of resistance that is horizontal. How big is this remnant? In the normal course of breeding for vertical resistance, selection for horizontal resistance is suspended and horizontal resistance is eroded. When the vertical resistance is lost, all that remains of the total resistance is a remnant of eroded horizontal resistance. The erstwhile resistant cultivar is now more susceptible than cultivars that were not bred for vertical resistance. This is the *vertifolia effect*. It is the erosion of horizontal resistance in the course of breeding for vertical resistance. The remnant of resistance, which is the topic of this chapter, is decreased. In the period of bust the farmer is worse off than if he had not used a cultivar with genes for vertical resistance.

Second, what happens when the vertical resistance has a nearly isogenic background? Vertical resistance is here considered against a background of uniform horizontal resistance. When vertical resistance is lost, is the loss incomplete? Does vertical resistance leave a remnant when it is defeated? This postulated remnant has been called *ghost resistance* in the literature.

Third, irrespective of whether the background of the vertical resistance is heterogeneous or nearly isogenic, does the pathogen in defeating the vertical resistance itself lose aggressiveness? Is the pathogen's victory pyrrhic? Is there a remnant of resistance which in reality is not a gain by the host but a loss by the pathogen? This seeming gain by the host has been called the *horizontal resistance equivalent*.

In determining resistance remnants the vertifolia effect is usually more important than ghost resistance or the horizontal resistance equivalent. Consider the resistance of the potato cultivar Pentland Dell to *Phytophthora infestans*. Pentland Dell combines the blight resistance genes *R1*, *R2*, and *R3* and had the highest vertical resistance to blight of any potato cultivar to achieve agronomic success. Soon after farmers started to plant Pentland Dell, virulent races of *P. infestans*, i.e., races with 1, 2, and 3 in their designations, became common where the cultivar was grown. In the presence of these races Pentland Dell is abnormally susceptible to blight both in the tubers and foliage. In its maincrop maturity class, Pentland Dell, in the presence of virulent races, is probably the most blight-susceptible potato cultivar in Britain. The resistance of Pentland Dell crashed in the course of a boom-and-bust cycle from the highest vertical resistance to the lowest horizontal resistance in its maturity class. Here was a vertifolia effect at its worst. By comparison, we must consider to be trivial any ghosts surviving the defeat of the resistance genes *R1*, *R2*, and *R3* or any reduced aggressiveness of the virulent races of the pathogen. For all practical purposes it was the vertifolia effect that decided how much horizontal resistance remained in Pentland Dell.

One notes too that the history of Pentland Dell and the vertifolia effect in general make nonsense of the suggestion that genes for horizontal resistance are simply defeated genes for vertical resistance. Were this suggestion true, Pentland Dell, in defeat, would still be horizontally the most resistant instead of probably the most susceptible maincrop variety in Britain.

8.2   THE VERTIFOLIA EFFECT

The vertifolia effect is defined as the erosion of horizontal resistance by plant breeders when their breeding material is protected from disease by vertical resistance. Consider some history of potato breeding in Europe.

In the 1920s potato breeders started to use the gene *R1* for resistance to *P. infestans*. The first discovery of virulence for gene *R1* was reported in Germany by Schick (1932). For two decades this virulence remained uncommon, because most farmers still grew potato cultivars without *R* genes; of 34 isolates of *P. infestans* collected by Frandsen (1956) from farmers' potato fields in northwestern Germany in the early 1950s only 1 was virulent for the gene *R1*. By that time many new potato cultivars with the gene *R1* had been released in Germany, and their cultivation was increasing fast. With this increase came an increase of virulence in *P. infestans* populations for this gene. Table 8.1 indicates the state of affairs in Germany in 1956; on August 3, the date most relevant to our analysis, 14.6% of the isolates were virulent for gene *R1*. Table 8.1 also shows that virulence for gene *R3* was still rare, the gene *R3* having been used by potato breeders after, and less commonly than, gene *R1*.

Against this background we can consider the data of Kirste (1958). The data are for blight in 12 late-maturing potato cultivars. Late-maturing cultivars are the most relevant because they are exposed to the blight epidemics that come in the latter part of summer and they are normally the most resistant. Kirste's experiments were carried out in 1955. It takes at least 10 years to breed, select, test, and release a new potato cultivar, and it can be accepted that the cultivars with *R* genes in Table 8.2 were developed at a time when virulence was rare and vertical resistance highly effective. They were effectively protected from blight at the time breeders selected them. From them one can determine how horizontal resistance to blight fared in breeding programs conducted in the virtual absence of blight.

TABLE 8.1

Occurrence of Various Races of
Phytophthora infestans in Germany in 1956[a]

Race	Frequency (%)		
	Aug. 3	Aug. 29	Oct. 5
(4)	85.4	64.6	58.5
(0)		5.4	3.2
(1)	10.4	21.2	27.4
(1, 4)	4.2	7.6	9.7
(1, 2)		1.2	
(1, 3, 4)			1.1

[a] Data of Schick *et al.* (1958b) for 209 isolates of *P. infestans* from potato varieties without *R* genes.

TABLE 8.2

Progress of an Epidemic of *Phytophthora infestans* on Various Late-Maturing Potato Varieties in Germany in 1955[a]

Variety	R genes[b]	Date of reaching blight rating[c]			
		1	2	3	4
Lerche	0	Aug. 3[d]	Aug. 7	Aug. 25	
Heimkehr	0	Aug. 3	Aug. 16	Aug. 28	
Ackersegen	0	Aug. 3	Aug. 18	Aug. 28	
Carmen	0	Aug. 7	Aug. 25	Sept. 4	
Ronda	0	Aug. 10	Aug. 16	Sept. 1	Sept. 4
Capella	0	Aug. 13	Aug. 16	Sept. 20[e]	
Ancilla	R1	Aug. 13	Aug. 16	Sept. 1	
Monika	R1	Aug. 13	Aug. 19	Sept. 1	
Adelheit	R1	Aug. 13	Aug. 22	Aug. 28	
Herkule	R1	Aug. 13	Aug. 19	Sept. 4	
Panther	R1	Aug. 13	Aug. 19	Aug. 25	
Vertifolia	R3, R4	Aug. 22		Aug. 28	Sept. 1

[a] From data of Kirste (1958).

[b] No R gene present = 0.

[c] Ratings: 1 = very mild attack, only occasional lesions found; 2 = mild attack, lesions on about one leaf per plant; 3 = medium infection, several leaves per plant attacked; 4 = all leaves attacked, but the plants generally still green.

[d] The date for the six varieties without R genes reflects horizontal resistance; the date for the six varieties with R genes reflects the combination of vertical and horizontal resistance. The very late recorded start, August 22, of blight in Vertifolia largely reflects the variety's very great vertical resistance in the mid 1950s, which in turn reflects the very low frequency at that time of virulence for gene R3 (see Table 8.1). The intervals between the dates of successive ratings are determined by horizontal resistance.

[e] Capella was still at stage 2 when last examined on September 20.

Table 8.2 shows that on an average it took 15.5 days for blight to increase from severity 1 to severity 3 in the six cultivars with R genes. But in the six cultivars without R genes it took as much as 26.7 days. (Even this figure is an underestimate because in the cultivar Capella severity 3 was never reached at all by the time the observations ended.) That is, the infection rate of blight in the cultivars with R genes was on an average 26.7/15.5 times as great as the rate in the cultivars without R genes. This difference is highly significant statistically.

The difference means that the cultivars without R genes (i.e., without vertical resistance) had considerably more horizontal resistance. The same weather affected both groups of cultivars. On an average, blight in the six cultivars without R genes reached intensity 1 between August 6 and 7 and intensity 3 on September 2. On an average, for blight in the six (including Vertifolia) with R genes, intensity 1 was reached 8 days later and intensity 3 was reached 3 days earlier than in the six without R genes. The period for the cultivars without R genes straddles the period for cultivars with R genes. The cultivars without R genes were exposed to all the blight weather to which the cultivars with R genes were exposed. One can, as an illustration, consider the extremes. In Capella, without an R gene, blight reached intensity 1 on August 13 and was still at intensity 2 when examinations were discontinued on September 20. In Vertifolia, with genes $R3$ and $R4$, blight started late because of the rarity of virulence for gene $R3$, but having started, it developed quickly. It reached intensity 1 on August 22, intensity 3 on August 28, and intensity 4 on September 1. Weather that allowed Vertifolia to become quickly blighted in the last third of August failed to bring about comparable blighting of Capella. The difference between the two cultivars was not in the weather, but in their horizontal resistance.

Vertical resistance masks weak horizontal resistance. How great the masking can be is illustrated by Tables 8.3 and 8.4. Table 8.3 shows how Schick *et al.* (1958a) assessed the resistance of the late-maturing potato varieties studied by Kirste (1958) and included in Table 8.2. In the varieties without R genes that Schick *et al.* classed as very resistant, blight took, on an average, 32 days to increase from severity 1 to severity 3 in Kirste's experiments. The corresponding figure for varieties with R genes was 16 days. For the varieties with somewhat less resistance the figures were 25 and 12 days. The classification of Schick *et al.* was far too lenient for varieties with R genes. They classified these varieties as highly resistant even though the infection rate

TABLE 8.3

Resistance to *Phytophthora infestans* of Late-Maturing Potato Varieties:
A Comparison between the Classification of Schick *et al.* (1958a)
and the Findings of Kirste (1958)

Resistance class of Schick *et al.*[a]	Days from blight severity 1 to severity 3[b]	
	Varieties with R genes	Varieties without R genes
1	16	32
2	12	25
3		22

[a] Very resistant = 1, very susceptible = 5.
[b] Averages from Kirste's data in Table 8.2.

TABLE 8.4

Resistance to Phytophthora infestans **of Late-Maturing Potato Varieties:
A Comparison between the Classification of Hogen Esch and Zingstra (1957)
and the Findings of Kirste (1958)**

	Days from blight severity 1 to severity 3[b]	
Resistance class of Hogen Esch and Zingstra[a]	Varieties with R genes	Varieties without R genes
9	16	
8	6	30
7		
6		24

[a] Immune = 10, very susceptible = 3.
[b] Averages from Kirste's data in Table 8.2.

in them was (in Kirste's experiment) twice as great as in varieties without R genes.

Schick et al. (1958a) did not allow for differences in the prevalence of the relevant races of P. infestans, i.e., in the initial inoculum from which blight developed in Kirste's plots. They made the common mistake of assuming that as long as some inoculum is present it does not greatly matter how much is present. For example, they rated both Capella and Vertifolia as very resistant. Capella is attacked by all races of P. infestans. Vertifolia is attacked only by races that in the mid 1950s were rare in the general population of P. infestans; there was an inevitable time lag needed for these races to catch up. Time lags are discussed in Chapter 13.

Before leaving the assessments of Schick et al. (1958a), it is worth interpolating that within each of the two groups of varieties, i.e., within the group with R genes and the group without R genes, their assessments agree with the results of Kirste. The lower they assessed the resistance of varieties, the shorter was the interval between blight severity 1 and 3 in Kirste's experiments. Thus, in varieties without R genes, classes 1 (most resistant), 2, and 3 corresponded with intervals of 32, 25, and 22 days, respectively.

Hogen Esch and Zingstra (1957) in the Netherlands also examined a large collection of potato varieties and assessed them for resistance to blight. In Table 8.4 their assessments are also compared with the findings of Kirste. Only three (Monika, Panther, and Vertifolia) of the six late-maturing varieties with R genes in Kirste's experiments appear in Hogen Esch and Zingstra's list. The table is therefore somewhat incomplete. Nevertheless, the difference between the varieties with R genes and those without them is statistically significant. Table 8.4 shows the same features as Table 8.3. Like Schick *et al.* (1958a), Hogen Esch and Zingstra also greatly overestimated the horizontal resistance of varieties with R genes. They too assessed Capella and Vertifolia as equally resistant.

In breeding new cultivars when the parent material, progenies, and selections are all protected from disease by vertical resistance, there is no check on the erosion of horizontal resistance. The erosion is great. All is well, provided that the vertical resistance remains stable. But if the vertical resistance fails through matching virulence, agriculture is left worse off than before.

High standards of agronomy or horticulture are incompatible with high horizontal resistance. That appears to be the disquieting message of the vertifolia effect. If it were only from the vertifolia effect that the message came, one could perhaps ignore it and seek other explanations. Unfortunately, there are other sources of the message, and in Chapter 11 the possibility is probed that high yields of grain or tubers are inimical to high horizontal resistance. The possibility cannot be swept under the rug.

8.3   GHOST RESISTANCE

A theory of ghost resistance was suggested by Riley (1973). The theory states that when an R gene for vertical resistance in the host plant is defeated by a matching virulence gene in the pathogen, a ghost of resistance survives. That is, a defeated R gene is better for the host plant than no R gene at all. The consensus has been against the existence of ghost resistance. Thus, Samborski and Dyck (1982) found that, judged by reaction types, a backcross line from Neepawa wheat with the leaf rust resistance genes $Lr11$ and $Lr30$ was defeated just as completely by virulent cultures of *Puccinia recondita* as was Neepawa itself. Nevertheless, the theory of ghost resistance must be judged on its merits, without prejudice from previous evidence.

Martin and Ellingboe (1976) found that the compatibility between the wheat powdery mildew resistance gene $Pm4$ and the corresponding virulence gene $p4$ in *Erysiphe graminis tritici* was not as great as that between the susceptibility gene $pm4$ and the avirulence gene $P4$. Defeated by virulence the gene $Pm4$ still left ghost resistance behind. Nass *et al.* (1981) worked with the same disease. Six near-isogenic lines of wheat, each with a different powdery mildew resistance gene, were evaluated for a residual effect when defeated by *E. graminis tritici* with virulence for all six resistance genes. The isolines with the genes $Pm3c$, $Pm4$, and MA demonstrated significant residual effects; even in defeat they were more resistant than the control susceptible wheat cultivar Chancellor. Nelson *et al.* (1982) repeated the evidence for genes *Pm3c* and *Pm4*; in defeat they left behind a ghost resistance measured as a reduction in sporulation and lesion numbers.

Anderson (1982) challenged the interpretation of Nass *et al.* (1981). Among other points he showed that their evidence could be explained by assuming linkage and genetic drift. In detail, the six resistance genes used by Nass *et al.* were *Pm2*, *Pm2+*, *Pm3c*, *Pm4*, *Pm5*, and *MA*. The isoline with gene *Pm2+* exhibited the very opposite of a ghost effect. Challenged by virulent *E. graminis tritici*, it had 76% more sporulating colonies than Chancellor in one experiment and 58% more in another. The lines with *Pm2* and *Pm5* had about the same number of sporulating colonies as Chancellor. The lines with *Pm3c*, *Pm4*, and *MA* had less. Anderson pointed out that this range of results in a quantitative trait is compatible with the hypothesis that genetic drift of quantitative resistance genes, unrelated to the *Pm* genes, occurred during the breeding of the near-isogenic lines.

Rowell (1981a,b, 1982) showed that wheat lines with the gene *SrTt1* had less stem rust when challenged by *P. graminis tritici* with putative virulence for this gene. Seedlings with the gene *SrTt1* gave a mesothetic (intermediate) reaction type. There seemed to be a preanthesis resistance and postanthesis loss of resistance much like that to be discussed in Chapter 15. To assess Rowell's evidence one must examine the basic assumption of the literature on ghost resistance.

8.4   THE EITHER/OR AVIRULENCE/VIRULENCE ERROR

It is a common assumption in the literature to assume the existence of only two alleles for avirulence/virulence. A genotype of a pathogen is, on this assumption, either virulent for a particular resistance gene or it is not. This assumption is wrong. Within a pathogen virulence is determined by multiple alleles that condition considerable variation. Pick your allele and you can pick the ghost. With variable virulence the supply of ghosts is unlimited. The assumption of a single allele for virulence is the very basis of the evidence for the ghost hypothesis, and the purpose of this section is to dispute it.

Watson and Luig (1968b) selected a culture of *P. graminis tritici* that produced gray-brown uredospores. The purpose of using specially colored spores was to insure against undetected contamination. The culture was avirulent for wheat plants with the gene *Sr11*; the reaction type with these plants was 0, indicating high resistance. Starting with a single spore they increased the culture on plants of a susceptible variety and then retested on plants with the gene *Sr11*. One pustule was found, giving an intermediate reaction type X=. This in turn was increased on plants of a susceptible variety and then again retested on plants with the gene *Sr11*. They then found a pustule with a mildly susceptible reaction 3. (The culture from it gave a 2+, 3c, and 3 reaction on the same leaf.) Wheat plants with the gene *Sr11* give a 4 reaction with fully virulent cultures, so here we have evidence for four reaction types, 0, 2=, 3, and 4 on plants with the gene *Sr11*. Four cultures with different reaction types would be recognized by their behavior on plants with gene *Sr11*.

Virulence for the gene *Sr9g* is very unstable, and progressive increases or decreases occur naturally (Luig, 1979). Indeed, progressive levels of virulence and avirulence now appear to be typical of most *Sr* loci in wheat (McIntosh, 1977). There are five known levels of interaction involving virulence and avirulence for the gene *Sr15* (McIntosh, 1977). In oats Luig and Baker (1973) noted that there were two reaction types (1 and 2+) representing resistance when the variety Rodney, with the gene *Pg4*, was inoculated with *P. graminis avenae*, and they ascribed this to multiple alleles for virulence in the pathogen.

Evidence for multiple alleles with the same recognition system also comes from experiments in which a culture was trained to become more pathogenic. Whereas in the experiments just considered the mutations occurred while the pathogen was multiplied on plants of a susceptible variety, training is effected on plants of a resistant variety. Reddick and Mills (1938) and de Bruijn (1951) trained *Phytophthora infestans* to become gradually more virulent. They inoculated plants of a resistant variety from which they managed to obtain a few spores. These were used to start a culture that was slightly more virulent. Plants of a resistant variety were successively inoculated, and successively more and more spores were obtained until finally the culture was fully virulent.

The gene-for-gene hypothesis must be elaborated to read that for every resistance gene in the host there is a corresponding and specific virulence gene in the pathogen with an indefinite number of alleles conditioning varying virulence.

To return to ghost resistance, there can be as many ghosts as there are virulence alleles. A selected culture means a selected ghost. One must beware of artifacts and accept that proof of a ghost in the laboratory is not proof of a ghost in the field.

The resistance gene *SrTt1* is rare or absent in the wheat fields of North America susceptible to stem rust (see Table 4.5). Virulence in *Puccinia graminis tritici* for the gene *SrTt1* comes not as a conqueror but by association with virulence for other resistance genes in the XYZ group. Full pathogenic fitness cannot therefore be assumed. In the eastern wheat belt of Australia, on the other hand, virulence for the gene *SrTt1* conquered the wheat cultivar Mengavi in the early 1960s (see Section 5.2 and Fig. 5.3). Mengavi has the gene *SrTt1*. Its destruction was swift, the virulence allele in the pathogen evidently fit, and there was no evidence for the existence of ghosts or record of mesothetic reaction types.

As Anderson (1982) has pointed out, if defeated resistance genes have residual effects against certain isolates of the pathogen, and if this is only because these isolates have suboptimal virulence alleles, then such residual effects would be likely to disappear under field conditions as more virulent strains became predominant through natural selection.

8.5   THE HORIZONTAL RESISTANCE EQUIVALENT

The horizontal resistance equivalent is a seeming quality of the host that in reality is a quality of the pathogen. A pathogen with virulence for resistance genes from each of the ABC and XYZ groups loses fitness (see Chapter 4); and a loss of fitness in the pathogen is the equivalent of a gain of resistance in the host. A loss of fitness could seemingly leave a remnant of resistance.

An order of importance must be kept in mind. A loss of fitness of the pathogen determines a horizontal resistance equivalent in the host. But much more importantly, it stabilizes or helps to stabilize vertical resistance in the host by suppressing superraces of the pathogen. It is much more important to keep vertical resistance booming than to lessen the shock of its busting. Practical plant breeders will aim at maintaining full vertical resistance rather than a remnant of this resistance.

9

Protein
Polymorphism and
Vertical Resistance

9.1 INTRODUCTION

Vertical resistance, insofar as it involves a gene-for-gene relation, is just another manifestation of protein polymorphism* in host and pathogen.

Specificity in host–pathogen relations resides in susceptibility. That is the clear, unequivocal message of Chapter 2. Specificity in susceptibility is the key to the understanding of what will now be discussed, and Chapter 2 should be reread if there is any doubt about it. The specificity of susceptibility is not only one of the pillars of the protein-for-protein hypothesis of gene-for-gene relations and vertical resistance; it also disposes of those theories of host–pathogen relations that miss the point.

The protein-for-protein hypothesis states that gene-for-gene recognition

---

* Genetic polymorphism is the occurrence together within a given population of two or more discontinuous variants in frequencies greater than can be maintained merely by recurrent mutation. In this sense it was defined by Ford (1945) and is generally used in modern protein studies. In the literature of gene-for-gene relations polymorphism has regrettably been associated with allelism. No such association is necessarily required by the definition, and the genetic background is left open. In the most familiar example of polymorphism, the occurrence of two sexes within a population, the genetic background is diverse and varies from group to group of animals, plants, and fungi.

is protein-for-protein recognition. In susceptibility, host protein and pathogen protein associate (polymerize) specifically. There are three pillars on which the hypothesis is built: Specificity in gene-for-gene systems is in susceptibility; the relevant molecules store variation massively; and susceptibility is endothermic in gene-for-gene systems.

9.2 MOLECULAR ASSOCIATION IN SPECIFIC SUSCEPTIBILITY

Specific susceptibility implies that in susceptibility there is specific molecular association within the range of chemical bonds.

Specific susceptibility requires a specific inducer produced by the pathogen and a specific receptor produced by the host. This requirement is evident in the diagonal of Fig. 2.1. In susceptibility, i.e., in disease, inducer and receptor associate, and it is with the thermodynamics of molecular *associations* that the protein-for-protein hypothesis is concerned.

9.3 MOLECULAR STORAGE OF MASSIVE VARIATION

The associated molecules are polycompounds capable of storing and recognizing massive variation.

Consider the three rust diseases of wheat. Wheat has at least 40 known *Sr* genes for resistance to *Puccinia graminis*, 35 known *Lr* genes for resistance to *P. recondita*, and 10 known *Yr* genes for resistance to *P. striiformis* (Priestley, 1978). Add to this genes for resistance to powdery mildew, smuts, bunts, and insects known or thought to be on a gene-for-gene basis, and one must allow in wheat for at least 90 genes for resistance and 90 for susceptibility to various diseases. (One need not quarrel over the exact number. There is some overlapping, e.g., genes *Sr15* and *Lr20* are probably identical. To offset this, new genes are being discovered every year.) There are thus potentially 2^{90} ($\sim 10^{27}$) phenotypes of wheat distinguishable by one or another pathogen or race of pathogen. This reflects massive *qualitative* variation in wheat.

So, too, there is massive qualitative variation in some pathogens. On the gene-for-gene hypothesis the 35 *Lr* genes in wheat mean 35 genes for virulence and 35 for avirulence in *P. recondita*. There are thus 2^{35} ($> 10^{10}$) phenotypes of *P. recondita* capable of being distinguished by appropriate lines of wheat. (The alleles for virulence discussed in Section 8.3 would bring the estimate still higher.)

The number 90 or, better, $90 \times 2 = 180$, prescribes the number of qualitatively different molecules that must be accounted for. These are the qualitatively different molecules in the host that pathogens must be able to recognize. Qualitative variation of this magnitude is possible only with polycompounds: polynucleotides (DNA, RNA), polypeptides (proteins), and polysaccharides. In these compounds variation in the order, arrangement, and number of the subunits permits variants to exist in almost endless permutations and combinations.

All heritable variation is stored in the DNA (the RNA viruses apart) and passes along in the sequence DNA $\rightarrow$ RNA $\rightarrow$ protein $\rightarrow$ polysaccharide. Where in this sequence do the molecules carrying the variation of the host and those carrying the variation of the pathogen associate and recognize one another? That, at the molecular level, is the central question of specific susceptibility.

9.4   ENDOTHERMIC SUSCEPTIBILITY

The relevant molecular associations in gene-for-gene disease are endothermic, which points to protein associations. DNA, RNA, and polysaccharides associate predominately by hydrogen bonding, proteins predominately by hydrophobic bonding. Hydrogen bonding is exothermic; hydrophobic bonding, endothermic. Susceptibility in gene-for-gene disease is endothermic. The evidence for this is considerable (Vanderplank, 1978, 1982); a fragment appears in the next section. The thermodynamic evidence is thus for protein associations.

9.5   INDIVIDUALITY IN TEMPERATURE RESPONSES

Individual temperature responses of genes confirm that specificity is determined by endothermic reactions.

Martens *et al.* (1967) found that the oat stem rust resistance genes *Pg3* and *4* were effective against *P. graminis avenae* only up to 20°C; genes *Pg8, 9, 13,* and *15* were variably effective at 25°C but ineffective at 30°C; and only genes *Pg1* and *2* were effective at 30°C, the highest temperature studied and almost the highest that could possibly have been studied, because 30°C is near the cardinal maximum temperature for oat stem rust. The change from resistance to susceptibility with rising temperature shows that susceptibility is endothermic.

The temperature response is thus seen to vary with the different *Pg* genes. This individuality is a piece of luck for the protein-for-protein hypothesis. Had there been no individuality, i.e., had all the *Pg* genes reacted similarly, a key to the hypothesis would have been missing. Had they all, like genes *Pg1* and 2, remained effective at the highest temperature, 30°C, there would have been no evidence that susceptibility is endothermic. Had they all, like genes *Pg3* and 4, been effective only up to 20°C, there would have been evidence that susceptibility is endothermic, but no evidence that the endothermic process is the process that determines specificity. The individuality of the temperature responses settles the issue. The individuality demonstrates not only that susceptibility is endothermic but also that the specificity of susceptibility is to be sought in the endothermic process.

Individual temperature responses occur with the *Sr*, *Lr*, *Yr*, and *Pm* genes in wheat and with some other genes in gene-for-gene systems.

9.6   VERTICAL RESISTANCE

Vertical resistance means that the host protein and pathogen protein fail to associate or, if they associate, fail to complete the development of the appropriate quaternary structure. What in Section 9.2 is described as the inducer of the susceptibility reaction (i.e., of pathogenesis) becomes the elicitor of defense reactions. The inducer of susceptibility represented by entries on the diagonal in Table 2.1 is also the elicitor of resistance represented by entries off the diagonal. The assumption here is that the pathogen releases the same product on or into the host cell, irrespective of whether the host is susceptible or resistant; the pathogen "discovers" whether the host is susceptible or resistant only after the release. This assumption accords with the evidence.

9.7   COPING WITH RECESSIVE RESISTANCE

Theory predicts and experience confirms that in gene-for-gene systems recessive resistance can be changed to dominant resistance by lowering the temperature. If at ordinary temperatures resistance is recessive, plant breeders can facilitate the selection of segregating progenies by the simple device of using lower temperatures when they test for resistance.

9.8   TEST TO DISTINGUISH VERTICAL FROM HORIZONTAL
PARTIAL RESISTANCE

How can one test in advance of field performance whether the partial resistance of a cultivar is horizontal or vertical? Fig. 9.1 suggests a test: If the optimum temperature for disease is higher in the resistant cultivar

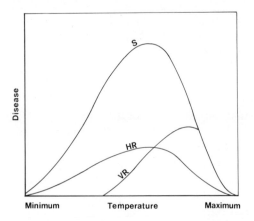

Fig. 9.1.   The effect of resistance on the optimal temperature for disease. S = susceptible variety; HR = variety with horizontal partial resistance; and VR = variety with vertical partial resistance. The test is inept if vertical resistance becomes ineffective at temperatures below the normal optimum.

than in a standard susceptible cultivar, the partial resistance is probably vertical. Disease must be measured in such a way that resistance is clearly demonstrated.

For vertical resistance the reasoning is based on endothermic susceptibility. For horizontal resistance the assumption that the optimum temperature for disease varies little with varying resistance is based on negative evidence. For example, there is a very wide range of horizontal resistance to *Phytophthora infestans* in potato cultivars, but blight forecasting systems use the same temperature criteria, irrespective of susceptibility.

The test for horizontal resistance, unlike the test for vertical resistance, is empirical. Mutation can affect the optimum temperature of enzyme activity. It is to be expected that variation in horizontal resistance will also affect it, but the effect is likely to be too small to invalidate the test. More work is needed on the relation between resistance and the optimum temperature for disease; meanwhile, we should be wary of any resistance that substantially shifts the optimum into higher temperature ranges.

9.9   THE GENE-FOR-GENE HYPOTHESIS

All resistance on a gene-for-gene basis is necessarily vertical. It is not known whether, conversely, all vertical resistance is on a gene-for-gene basis. However, the most conspicuous examples of vertical resistance seem to be covered by the gene-for-gene hypothesis, so a brief discussion is relevant.

In his work on flax rust, Flor (1942) was the first to study the genetics of both members of a host–pathogen system. From his experiments he concluded that for each gene determining resistance in flax (*Linum usitatissimum*) there was a specific and related gene determining pathogenicity in the rust fungus (*Melampsora lini*). In flax varieties possessing one gene for resistance to the avirulent pathogen, pathogenicity in a virulent race was conditioned by one gene in the fungus. In flax varieties possessing two, three, or four genes for resistance, pathogenicity was conditioned by two, three, or four genes in the fungus. The hypothesis, that for each resistance gene in the host there is a matching or reciprocal gene for pathogenicity in the fungus, is the simplest that fits these facts. The range of pathogenicity of a race of *M. lini* is determined by pathogenic factors specific for each resistance factor possessed by flax.

TABLE 9.1

Parasite–Host System for Which a Gene-for-Gene Relationship Has Been Shown or Suggested

Classification	Parasite–Host System
Viruses	Tobacco mosiac virus–*Lycopersicon*
	Spotted wilt virus–*Lycopersicon*
	Potato virus X–*Solanum*
Bacteria	*Xanthomonas malvacearum–Gossypium*
	Rhizobium–Leguminoseae
Phycomycetes	*Phytophthora infestans–Solanum*
	Synchytrium endobioticum–Solanum
Ascomycetes	*Erysiphe graminis hordei–Hordeum*
	E. graminis tritici–Triticum
	Venturia inaequalis–Malus
Basidiomycetes	*Melampsora lini–Linum*
	Hemileia vastatrix–Coffea
	Puccinia graminis avenae–Avena
	P. graminis tritici–Triticum
	P. helianthi–Helianthus
	P. recondita–Triticum
	P. sorghi–Zea
	P. striiformis–Triticum
	Ustilago avenae–Avena
	U. hordei–Hordeum
	U. tritici–Triticum
	Tilletia caries–Triticum
	T. contraversa–Triticum
	T. foetida–Triticum
Deuteromycetes	*Fulvia fulva–Lycopersicon*
Nematodes	*Globodera rostochiensis–Solanum*
Insects	*Mayetiola destructor–Triticum*
Angiosperms	*Orobanche–Helianthus*

The hypothesis, that for every resistance gene in the host there is in disease a corresponding virulence gene in the pathogen, is an experimentally based principle. In the flax rust system 26 resistance genes in the host have been matched with 26 virulence genes in the pathogen; for this system the principle is strongly based on experimental fact. In other systems the strength of the evidence varies from very high, particularly in the wheat–*Puccinia graminis* system, to rather poor. Table 9.1 lists various systems for which a gene-for-gene relationship has been demonstrated or suggested.

The systems listed in Table 9.1 are probably only a very small sample of reality. For example, only one species of powdery mildew, *Erysiphe graminis*, occurs in the list out of many known; one must suspect that its unique feature is that a unique amount of genetic attention has been given to powdery mildew of wheat and barley. In general one expects to hear of gene-for-gene systems only where genetic research has been concentrated with the aim of developing resistant cultivars. Also, there must be resistance genes readily transferable. *Solanum tuberosum* has no *R* genes for resistance to *Phytophthora infestans;* a gene-for-gene system could be suggested only after *R* genes were introduced from *S. demissum* and only because *S. tuberosum* and *S. demissum* hybridize to give progeny that potato breeders thought worth studying.

The gene-for-gene hypothesis has been reviewed by Flor (1971) himself and discussed by several others including Day (1974), Sidhu (1975), Person and Ebba (1975), and Vanderplank (1982).

10

Genes for Susceptibility

This chapter, like Chapter 9, deals with susceptibility, but with a difference. Chapter 9 deals with susceptibility when genes for resistance in the host are matched by genes for virulence in the pathogen. The present chapter deals mainly with susceptibility when genes for susceptibility in the host are matched by genes for avirulence (or virulence) in the pathogen. The previous chapter deals with R genes; this chapter with r genes.

What is an r gene for susceptibility? It is a plant gene, not just a host plant gene. It has a primary role in the healthy plant that has nothing to do with parasitism. It is not primarily a gene of invitation to the parasite. Its role in parasitism is secondary, forced upon it by the parasite which turns the plant's processes to its own ends.

What then is the difference between the two host–pathogen combinations that make the host susceptible: the resistant host–virulent pathogen combination, the topic of the previous chapter, and the susceptible host–avirulent (or virulent) pathogen combination, with which we must now deal? In both combinations the process of disease is essentially the same. It is the process of feeding the parasite. Phenotypically (if one ignores questionable ghost resistance), the combinations do not differ. An observer of a diseased field, if he did not have prior knowledge of the genetics of the host plants, would be unable to tell whether the host plants lacked resistance genes or the pathogen had virulence genes. There is evidently a reciprocal arrangement which, in terms of the protein-for-protein hypothesis, we see as reciprocal mutations for hydrophobicity in the amino acid residues of the proteins. Compared with the resistant host–virulent pathogen system, the susceptible host–avirulent pathogen system involves a more hydrophobic host protein and a less hydrophobic pathogen protein.

It is to be understood that, to avoid repetitive qualifications, this chapter dealts with the same sort of diseases as the previous chapter, i.e., the sort of diseases listed in Table 9.1.

10.2   VAVILOV'S RULE

The gene centers of cultivated plants are the best places to find resistance to disease; disease resistance should be sought in the homelands of the hosts and their parasites. That is Vavilov's Rule (Vavilov, 1949). Evidence for the rule has been discussed and reviewed by Anikster and Wahl (1979), Flor (1971), Harlan (1961), Leppik (1966, 1970), Reichert (1958), Rudorf (1959), and Zhukovsky (1961); and the rule has been widely supported.

It is, however, not with the rule but its corollary that this chapter is directly concerned.

10.3   THE COROLLARY OF VAVILOV'S RULE

Genes for susceptibility are essential plant genes and, in general, make for greater fitness than resistance genes. That is the corollary to Vavilov's Rule.

By Vavilov's Rule, gene centers of cultivated plants are richer in resistance than elsewhere. Therefore, as plants left their gene centers they preferentially took with them genes for susceptibility. Genes for susceptibility, one must infer, are not only useful to the plant but also, on a definition of fitness as ability to leave progeny, more able to confer fitness.

The tendency towards resistance at the center and susceptibility elsewhere is manifest in the cultivars we use. Despite many decades of breeding for resistance to disease, genes for susceptibility still predominate numerically. There are 35 known *Lr* genes in wheat for resistance to *Puccinia recondita*, but few wheat cultivars have as many as 5 of them, as against 30 or more *lr* genes for susceptibility. There are about 40 known *Sr* genes in wheat for resistance to *Puccinia graminis*, but few wheat cultivars have as many as 6 of them, as against 34 or more *sr* genes for susceptibility. In the potato there are 11 known *R* genes for resistance to *Phytophthora infestans*, but the most important potato cultivars in the United States have none of them. Even the cultivar Kennebec, the most widely grown resistant cultivar, is only simplex for gene *R1*; it is triplex for *r1* and homozygous for susceptibility in the remaining 10 loci. Evidently, there are no advantages of heterosis to outweigh the potato's preference for susceptibility to blight. There is nothing exceptional about these examples. What has been said for wheat and *Puccinia recondita*, wheat and *P. graminis*, and potato and *Phytophthora infestans* could in essence have been said about any of the other host–pathogen combinations in Table 9.1. More than half a century of active plant breeding for resistance has failed to break the plant's association with what we call genes for susceptibility. Susceptibility as a phenotypic process has no virtue for the plant. The genes must therefore have some other role obscured by the term gene for susceptibility. They are useful plant genes; and one reaches this conclusion irrespective of whether one takes the susceptibility gene to be the original and the resistance gene the mutant, or vice versa.

10.4   SOME MISINTERPRETED EVIDENCE

Loegering and Sears (1981) suggested that alleles for susceptibility might be nonfunctional DNA sequences. This suggestion is at variance with the evidence of the previous section and needs discussion.

Loegering and Sears found that aneuploid lines of wheat lacking either genes *Sr6*, *Sr8*, *Sr9a*, or *Sr11* were susceptible to cultures of *Puccinia graminis tritici* when the corresponding euploid lines were resistant. The absence of a locus due to aneuploidy was the apparent equivalent of an allele for susceptibility, whence their suggestion that the allele might be a nonfunctional DNA sequence. This suggestion goes beyond their evidence, because it would be logical only if the locus concerned were the only locus involved in the wheat–*P. graminis tritici* system. If, as in Section 9.3, it is accepted that in wheat there are 40 or more genes for resistance/susceptibility to *P. graminis tritici*, their argument falls away. A simpler suggestion is that in the susceptible aneuploid lines the pathogen was living on the products of the remaining 39 or more loci. A 1/40 or 2.5% loss of food would have been undetectable in the experiments of Loegering and Sears.

10.5   BIOTROPHIC SEMIBENIGN INFECTION

The gene-for-gene hypothesis is linked to biotrophy. Of the pathogens listed in Table 9.1 some, like the viruses and rust and mildew fungi, are wholly or mainly biotrophic. Others, like *Phytophthora infestans*, *Venturia inaequalis*, and *Xanthomonas malvacearum*, are hemibiotrophs; they infect biotrophically at the start and then change to necrotrophy, and it is their biotrophic phase that concerns this chapter.

Biotrophy in plant disease implies the coexistence of pathogen and living host cell. Harm to the host cell is often small. The host cell nucleus stays alive. This semibenign interaction of host and pathogen has long been known in rust diseases in literature going back to Ward (1890, 1902, 1904). Mains (1917), writing about *Puccinia sorghi* in maize and *P. coronata* in oats, commented that although the host cells in the leaf are invaded by the haustoria of the fungus, the host cells suffer little apparent harm until long afterwards. Mycelium enters the intercellular spaces and haustoria invade the adjacent cells, yet the invaded cells retain the characteristics of cells of healthy tissue. Allen (1926) reported a large degree of compatibility between wheat and *P. recondita*; even 16 days after infection only 1% of the host cells, at most, were dead. Chester (1946), also discussing wheat leaf rust, found that each pustule lived to produce some 2000 spores per day for about 2 weeks. In maize infected with *P. polysora*, each pustule produces 1000–2000 spores a day for 18–20 days (Anonymous, 1958; Cammack, 1961). Durrell and Parker (1920) kept pustules of *P. coronata* alive on oat leaves for a month, all the time producing fresh crops of spores. Asai (1960) found that spore production by pustules of *P. graminis* in wheat remained stable for at least 10 days. All this production is in pustules showing little or no necrosis.

These observations about cereal rust diseases illustrate a general feature of all biotrophy: The host cell must stay alive and function efficiently; there must be no killing of the goose that lays the golden eggs. Normal processes like respiration must not be disrupted.

This brings us to the basic paradox. Genes for susceptibility are normal plant genes (Section 10.3), but in biotrophy normal processes must not be disrupted. The solution of the paradox that suggests itself is that the genes for susceptibility are genes that are needed to cope with stress: genes normally present but not normally used. Climatic stress is an unlikely candidate because climatic stresses are too variable. Biotrophy is found in climates wet or dry, hot or cold, sunny or shady. Wounding (injury) is a more likely candidate. Any plant cell is liable to injury, from wind, hail, trampling, feeding insects, grazing animals, and the like. Uninjured, a cell needs no repair; injured, it needs repair in a hurry. In plants, genes for repair are genes abnormally needed but normally to spare.

10.6   WOUNDS AND INFECTIONS: ROLE OF PEROXIDASE

The solution, that genes for susceptibility are primarily plant genes for wound repair, has genetic support. With genes coding for proteins, repetition, including duplication, is probably restricted to proteins needed abundantly in abnormal, critical situations. This was Price's (1976) conclusion from a survey of the DNA content of higher plants. This fits the solution. Genes for susceptibility are abundantly replicated. In wheat there are 40 or more for susceptibility to stem rust, 35 or more for susceptibility to leaf rust, and many others. And wounding is unquestionably an abnormal, critical situation.

The solution also has anatomical, physiological, and biochemical support, in the similarity between the host cell's responses to wounding and infection. There is a large body of literature about this. Our special concern is with proteins (for reasons clear in the previous chapter), and one protein, peroxidase, features largely in the literature common to both wounding and infection. From information now available it is the most likely candidate. Much of what will be said applies to other proteins; but for convenience we concentrate on peroxidase and its special features.

Peroxidase polymorphs (isoperoxidases, peroxidase isozymes) meet the requirement of being abundant and widely determined at different loci. Hoyle (1977) demonstrated by isoelectric focusing as many as 40 different bands of isoperoxidase in commercial preparations of horseradish peroxidase. Since only a minority of isozymes are detectable, the real number is probably in line with the requirements of the protein-for-protein hypothesis. Scandalios (1974) studied maize. He found 10 electrophoretically detectable zones of peroxidase acitivity. Polymorphism at these 10 regions is determined by allelic variation at each of 10 distinct loci. Scandalios concluded that gene duplication followed by mutation at the duplicated loci is a likely mechanism that can lead to the type of isozyme multiplicity that he observed. Again, we accept that for technical reasons Scandalios' observations did not reveal the multiplicity fully.

The number of loci is crucial to the hypothesis that genes for susceptibility (or resistance) are genes for wound repair, and the number indirectly tests the hypothesis. Given that in plants there are many genes for susceptibility or resistance (35 or more in wheat for susceptibility or resistance to *P. recondita*, 40 or more for susceptibility or resistance to *P. graminis*, and so on) it is a likely inference that plants have hundreds, perhaps thousands, of genes involved in the crash process of wound repair. Advances in the technique of analyzing protein polymorphs should soon allow this inference to be tested. The inference assumes that genes for susceptibility or resistance to different diseases are mostly independent of each other, although there are known examples of overlapping, e.g., in wheat genes, *Sr15* and *Lr20* are probably identical (McIntosh, 1973). Botanical research may yet elucidate a problem in plant pathology, and reciprocally plant pathology research may yet elucidate a problem in botany.

10.7   ULTRASTRUCTURAL EVIDENCE

In line with the protein-for-protein hypothesis we accept that a pathogen, be it a fungus, bacterium, nematode, or fly larva, makes its presence felt, and that the host cell responds as it would to other irritations, injuries, or wounds.

Some detail is given by Mendgen (1975) for compatible infections of *Phaseolus vulgaris* by *Uromyces phaseoli* in an ultrastructural demonstration of peroxidase activity after the haustorium of the fungus had penetrated the host cell. All the Golgi bodies that were found around haustoria contained peroxidase activity. They secreted vesicles that embodied peroxidase, and "the vesicles were obviously secreted towards the partly invaginated sheath of the young haustorium." Around the haustorium and especially near the Golgi bodies, peroxidase activity was seen on the rough endoplasmic reticulum. Here we have visual evidence (at magnifications of 23,600–40,500) of the host cell delivering protein to the pathogen's doorstep.

It is likely inference from the protein-for-protein hypothesis that proteolytic compartments (lysosomes) of the pathogen should be clustered at or near the haustorium's surface opposite to where Mendgen saw the host delivering protein. So far the difficulty of studying lytic compartments has defeated anatomists.

10.8   PROTEIN POLYMERIZATION

The protein-for-protein hypothesis is about how the pathogen gets some or most of its nitrogenous food. The host plant delivers protein. The pathogen attaches it by specific polymerization at or near the interface and carries it back, possibly along a gradient of thermodynamic potential to a thermodynamic sink (Vanderplank, 1982). The distance the polymer needs to travel across the interface is short, less than a micron, and the life of the polymer is perhaps measured in minutes rather than hours.

The specificity of protein polymerization is precise, and the protein-for-protein system has a counterpart in another protein-for-protein system of renowned precision: the protein antigen–antibody system.

Protein polymerization has to do not only with food transport and precise specificity but probably also with the derepression of appropriate genes. Only wanted genes are derepressed. To say that genes for wound repair are unused in the healthy, uninjured plant is to say that the genes are normally repressed. The question arises, how does *Puccinia graminis*, for example, derepress wanted genes in wheat for susceptibility to stem rust, leaving repressed the unwanted genes for susceptibility to other diseases? Alternatively, does the injury we call infection derepress all the genes for wound repair, with the unwanted genes quickly re-repressed by the plant's own autoregulatory process, as the products of the unwanted genes begin to accumulate?

In the pathogen the gene for avirulence (or virulence) is repressed in the young sporeling and derepressed later, presumably by the host's matching gene for susceptibility (or resistance) or its product. The mechanism is unknown.

Hydrophobicity is what links protein polymerization with genes and indirectly, polymorphism. To polymerize, proteins must be adequately hydrophobic (adequacy depending on temperature, pH, ionic strength, and solutes in the solvent). Hydrophobicity resides in the amino acid residues, and they link with the genetic code. Of the 20 essential amino acids, tryptophan (the most hydrophobic), tyrosine, phenylalanine, methionine, leucine, isoleucine, proline, and valine have hydrophobic residues, the methyl group in alanine is practically neutral, and glycine has no side chain. The rest have hydrophilic residues: serine, threonine, cysteine, aspartic acid, glutamic acid, lysine, arginine, asparagine, glutamine, and histidine. Most mutations (missense mutations) come through to the amino acids; most proteins have many, usually hundreds of, amino acid residues; variation in protein hydrophobicity is practically endless; and by the nature of the genetic code variation is reversible.

10.9   THE PATHOGEN'S PROTEIN

What sort of a protein do the pathogen's genes for avirulence (or virulence) code for? Almost certainly, it is not a peroxidase, because peroxidases in plants are known only as monomers. Two possibilities come to mind.

First, it is possible that the relevant protein of the pathogen has an unusually high proportion of hydrophobic amino acid residues. The relevant protein, so far as the evidence goes, is primarily a fishing line; the more hydrophobic it is, the better it can hook. (Protein polymerization includes heterogeneous associations. Protein polymers are associations of subunits without any implication that the subunits are homogeneous.)

Second, it is possible that polymerization is preceded by deglycosylation if the protein is a peroxidase. The sugar moiety must be removed at one time or another in the process of catabolism, and there seems to be no *a priori* reason why it should not be removed before polymerization. This would additionally give a source of energy at the interface.

10.10   THEORY OF RECIPROCAL MUTATION

Mutation from susceptibility to resistance in the host is a mutation to less hydrophobicity of the relevant protein. By less hydrophobicity we mean that the mutant amino acid residues are more hydrophilic or less hydrophobic. Mutation from avirulence to virulence in the pathogen is a reciprocal mutation to greater hydrophobicity of the relevant protein. By greater hydrophobicity we mean that the mutant amino acid residues are more hydrophobic or less hydrophilic. This is the molecular theory of boom-and-bust cycles.

Ability to explain reciprocal mutation in host and pathogen is essential in any hypothesis of gene-for-gene relations or of vertical resistance; and it is a strength of the protein-for-protein hypothesis that it does this, simply and directly. In its simplicity and directness the protein-for-protein hypothesis stands alone.

11
Sink-Induced Loss
of Resistance

11.1 INTRODUCTION

Horsfall (1975) recalled what he saw in tomato fields in the 1930s. Occasional plants were sterile, produced no fruit, and were highly resistant to *Alternaria solani*. They stood starkly alone as green islands in a sea of brown, defoliated plants. He produced similar resistant plants artificially by pinching off the blossom. He tried the reverse experiment, leaving the fruit but removing some of the foliage. The plants were highly susceptible. The susceptibility of the foliage could be manipulated at will by varying the amount of fruit the foliage had to support. These observations were recorded by Horsfall and Heuberger (1942). Rowell (1953) showed that the fruits remove sugar from the leaves, making them more susceptible to *A. solani*. This was the start of Horsfall's concept of low-sugar disease.

In current language, the fruit is a sink. Tomatoes have a sink-induced susceptibility to blight caused by *A. solani*.

While Horsfall was working on tomato blight Holbert *et al.* (1935) were making parallel observations with Diplodia stalk rot of maize. Their conclusions were practically identical. If the ear was removed, the stalks were resistant. If leaves were removed or damaged by disease or insects, the stalks were susceptible. Like Horsfall, Holbert *et al.* explained the result in terms of carbohydrate depletion. The ear was a sink that deprived the stalk of carbohydrate and made it susceptible to Diplodia rot.

Another crop with an old history of defruiting research is cotton. Eaton and Rigler (1946) and Batson *et al.* (1970) found that removing the bolls increased resistance to root rot caused by *Phymatotrichum omnivorum*. They too explained the effect of defruiting in terms of more carbohydrates available to the root bark. Eaton and Ergle (1953) and Bell (1973) found that removing the bolls increased resistance to Verticillium wilt, and Adkisson (1954) noted that Verticillium wilt appeared sooner in plants having heavy boll loads. Here again the evidence is consistent. Plants without a boll sink are more resistant to *Phymatotrichum* and *Verticillium*, and a large sink causes resistance to be lost.

Sink-induced susceptibility poses a major problem in plant breeding. Plant breeders inevitably aim at higher yields of grain, fruit, tubers, roots, and (in floriculture) flowers; higher yields mean bigger sinks; and bigger sinks (in many but not all diseases) mean lost resistance. Conversely, when plant pathologists set out to select for higher resistance to disease they are likely to select unconsciously for smaller sinks and crop varieties unacceptable to farmers because of poor yields.

This chapter probes the problem broadly, using stalk rot of maize as a vehicle of discussion.

11.2   STALK ROT OF MAIZE

Stalk rots occur worldwide in maize. Sink-induced susceptibility is known or has been suggested for stalk rots caused by *Colletotrichum graminicola*, *Diplodia maydis*, *Gibberella zeae* (*Fusarium roseum*, f.sp. *cerealis*), *Fusarium moniliforme* (*Gibberella moniliformis*), *F. moniliforme* var. *subglutinans*, and *Macrophomina phaseolina* (*Sclerotium bataticola*). These organisms attack plants approaching maturity. *Pythium aphanidermatum* and *Erwinia chrysanthemi zeae* are apart; they often attack plants before flowering and are not known to be affected by the size of the sink. Pythium and bacterial stalk rots are excluded from the discussion.

The fungi *Colletotrichum*, *Diplodia*, *Gibberella*, *Fusarium*, and *Macrophomina* occur widely, almost ubiquitously. They commonly occur together, although specific environments favor one or the other. With the exception of *Colletotrichum* they seem unable to damage vigorous young tissue but are confined to attacking plants nearing maturity. They are all necrotrophs, being nourished by breaking down the cellulose and lignin of old stalks.

Inoculum is apparently not a limiting factor, nor is the identity of the pathogen. Dodd (1980a) found that at the end of the season there was more or less the same amount of rotting regardless of whether the stalks were artificially inoculated with *Colletotrichum*, *Diplodia*, *Gibberella*, or soil or whether they were left uninoculated artificially. The host plant is an important variable. Some genotypes are more resistant than others, and there is a very strong ontogenic effect, susceptibility increasing sharply with the age of the plant. Environment is equally important. Stalk rotting is increased by an early environment that favors the setting of kernels, followed by conditions that are unfavorable. Unfavorable conditions include crowded plants, lack of sunshine, damage to the foliage by hail, leaf diseases, early frost, drought, injury by stalk-boring and root-boring insects, and a low K:N ratio.

The general concept of a source–sink balance in maize stalk rotting goes back to Holbert et al. (1935) and De Turk et al. (1937). The source of carbohydrate is the foliage and stalks. The sink is the developing grain. If the source cannot fully cope with the sink, the stalks are depleted of carbohydrate and predisposed to rots. In turn, the concept goes back to the work of Loomis (1934) on maize. In conditions of stress that restrict photosynthesis the amount of carbohydrate is insufficient for all demands. The demands of the developing grain are met first, and the level of sugar in the stalk drops.

A reduced carbohydrate source or an increased sink means more stalk rot. Consider first a reduced source. Holbert et al. (1935) found that the stalks of plants were more likely to rot if the leaves were injured by bacterial leaf blight in the second half of August or by second-brood clinch bugs which feed largely on the carbohydrates they remove from the lower parts of maize plants. They also found that artificial defoliation increased stalk rotting following both natural infection and hypodermic injection of a spore suspension of *Dipolodia maydis* while the leaves were still green. During the last 10 days of August they removed 25–35% of the tip end of each leaf by clipping, with results given in Table 11.1. Later, Wysong and Hooker (1966) showed that clipping reduces the concentration of soluble solids in the plant, and they also related this to increased rotting by *D. maydis*. Dodd (1980a) varied the procedure somewhat and reduced leaf area by cutting the stalks at various nodes above the ear. The stalks were cut 3 weeks after silking, and stalk rotting was recorded 8 weeks later. When the stalks were cut at the first or second node above the ear, 100% of the stalks rotted. When the cut was at the third node above the ear, the percentage was 85%, at the fourth node 76%, at the fifth node 70%, at the sixth node 37%, at the seventh node 27%, and with no cutting, i.e., with full foliage, 10%. The greater susceptibility to stalk rot in plants after the leaves have been attacked by disease has often been reported. Fajemisin and Hooker (1974) studied the effect of infection of the leaves by *Helminthosporium maydis* races T and O and *H. turcicum* on rotting by *D. maydis*. Plants with severe infection by the leaf blights developed more stalk rot than plants with low leaf blight infection, and the amount of blight damage was more important than the identity of the pathogen. Dodd

TABLE 11.1

Percentage of Maize Plants at Maturity
with Stalks Conspicuously Rotted
in Plant Populations with Leaves
Partially Clipped or Not Clipped[a]

Year	Clipped	Not clipped
1932	68	24
1933	50	11
1934	50	5

[a] Determined from a graph of Holbert et al.
(1935). The leaves were clipped during the last
10 days of August, from 25–35% of the tip end
of each leaf being removed. The pathogen was
Diplodia maydis.

(1980a) inoculated blight-susceptible and blight-resistant hybrids with
H. turcicum at the seven-leaf stage. The hybrids were near-isogenic and
differed essentially only in the absence or presence of the resistance gene *Ht*.
The blight-susceptible hybrids developed rot in 93% of the stalks, the
blight-resistant lines in only 1.7% of the stalks. Others have found that a
reduced rate of photosynthesis is as important as reduced leaf area. Cloud
cover and the shade that goes with high population density predispose
plants to stalk rot.

Consider now the sink. If the ear is removed, resistance to stalk rot
increases; if more grain is set, resistance decreases. Holbert *et al.* (1935)
recorded that in 1931 tissues of certain inbred lines not permitted to set
grain were invaded less rapidly and to a lesser extent following injection of a
spore suspension of *D. maydis* than were stalk tissues of comparable plants
of the same inbreds allowed to fruit normally. The inoculations were made
in the pith tissues of the third and fourth internodes above the soil line 3
weeks after pollination. Messiaen (1957) found that removing ears increased
resistance to *Gibberella zeae* and *Colletotrichum graminicola*. Dodd (1980a)
decapitated plants one node below the ear node; the ear-free remainders
developed no stalk rot, whereas plants decapitated one node above the ear all
developed stalk rot. Increasing the grain set increases the incidence of stalk
rot. It has been commonly observed that two-eared plants in an ordinarily
single-eared inbred or hybrid are specially susceptible to stalk rot. Dodd
(1980b) found that stalk-rotted plants have 10–19% more kernels than
adjacent healthy plants with no obvious differences in stress (see Table 11.2).
The apparent anomaly that stalk rotting is associated with a greater number
of kernels but a substantially smaller weight of kernels is readily explained by

TABLE 11.2

**Number of Kernels on Maize Plants with Rotted Stalks Compared
with the Number on Adjacent Plants with Healthy Stalks**[a]

Year	Hybrids (no.)	Pairs (no.)	Number of kernels on:		
			Rotted stalks	Healthy stalks	Diff.[b]
1976	40	112	561.9	495.0	66.9
1978	30	65	647.6	586.8	60.8

[a] From Dodd (1980b), with permission of the American Phytopathological Society.

[b] Significance level of $P = .001$ using Student's t test.

the fact that stalk rot hits plants while the grain is filling out. Koehler (1960) made an observation central to the theme of this chapter. In seasons and localities in which the incidence of stalk rot was low, he found that the maize hybrids known to be the most susceptible to stalk rot were the highest yielding. Inherent productivity and resistance were opposites.

11.3   SUGAR AND RESISTANCE TO MAIZE STALK ROT

Ever since maize stalk rot has been associated with ears as sinks it has been associated with carbohydrates in general and sugars in particular. As has already been noted, Holbert et al. (1935) and De Turk et al. (1937) ascribed stalk rot to carbohydrate depletion. Craig and Hooker (1961) followed the concentration of sucrose and reducing sugars in susceptible and in resistant inbreds; the pathogen was D. maydis. In susceptible inbreds there was a downward trend in reducing sugar concentration after silking; sucrose also decreased, but later and less consistently. In resistant inbreds, except WFG, the sucrose concentration increased after silking, and the reducing sugar concentration remained relatively stable. In WFG there was an increase of reducing sugar after silking, sucrose remaining relatively stable. In summary, total sugar decreased after silking in susceptible inbreds and increased in resistant inbreds. Mortimore and Ward (1964), studying natural infections, found that high levels of soluble sugars in the pith at physiological maturity are associated with resistance to stalk rot. Hybrids resistant to the disease had a higher sugar content than susceptible hybrids when grown under recommended cultural practices. Treatments that predisposed plants to stalk rot, namely, high population density and late defoliation, depleted the sugar content. Correspondingly, treatments that increased resistance to stalk rot, namely, the prevention of kernel development and low population density, resulted in a maintenance or increase of sugars in the pith. When sugars in a susceptible hybrid were maintained artificially at a level just as high as in resistant hybrids, stalk rot never occurred. Mortimore and Ward were able to put the matter quantitatively: The number of stalk-rotted plants never exceeded 10% unless the total sugar fell below 20% of dry weight in the pith of the lower internode at physiological maturity (represented by a kernel moisture of 35–37%).

There have been various theories about the effect of sugar on stalk rot. The latest is that of Dodd (1980a,b). The central theme is the photosynthetic stress–translocation balance. According to this theory, lower stalk tissues are decayed by several microorganisms as the tissues lose their metabolically dependent defense system because of carbohydrate deficiency. This deficiency occurs when photosynthesis is unable to cope with the drain of carbohydrates being translocated to the developing grain. Every plant needs carbohydrates to fill kernels. The needs are not static. Genetic consititution and the pre-flowering environment determine the number of kernels a plant sets. The number is variable. On the other hand, the rate at which carbohydrate moves to each kernel is fairly constant for 10–50 days after pollination. The daily rate per kernel varies with the genotype, but environment has little effect. Kernel number is thus the main variable among genetically identical plants that affects total translocation of carbohydrates to grain. Normally, about 20% of the grain weight comes from carbohydrate stored in the stalk. But if too many kernels are set or photosynthesis is reduced, stalks are depleted and become easily attacked by rotting organisms.

11.4 ECOLOGICAL TOPICS

Plant pathologists often laud the ecological state of plants in nature. Plants are in balance with their pathogens, and severe epidemics of disease are rare. Modern farming is blamed for most plant disease. Monoculture allows diseases to develop to an extent not possible in mixed plant communities. Genetic uniformity in crops allows virulent races of pathogens to sweep unhampered through fields. All this is true, but it omits one of the chief distinctions of modern farming. Modern farmers grow high-yielding cultivars and hybrids with bigger sinks.

Sink-induced loss of resistance is a price we must pay for higher yields. Our yield trials and studies of combining ability are aimed at producing maximum yields. Maximum yield is a legitimate aim, and developed agriculture can cope with its side effects. Fungicides can cope with diseases like Alternaria blight of tomatoes. Independent genes for resistance can be introduced to help balance sink-induced losses. In countries where agricultural systems are highly developed and sophisticated, higher yields can be pursued in the knowledge that side effects can be coped with or lived with.

It is when we come to consider farming in much of the tropics and subtropics that we should take note of sink-induced loss of resistance. Expatriate plant breeders come from temperate climates with the aim of raising yields. The aim is entirely meritorious. But there is a danger of sacrificing dependability for high yield. This is a danger discussed by Harlan (1972, 1980). He discussed it in relation to genetic uniformity. We see higher yields also as a danger in themselves: a danger of bigger sinks and lost resistance.

Landrace populations support much of subsistence agriculture. They are seldom outstanding in yield. But they are dependable, and in subsistence agriculture dependability is paramount. Resistance to epidemic disease contributes largely to dependability. The scope for raising yields judged by performance in experimental plots is great. Perhaps as great is the scope for epidemics of disease when new, high-yielding lines are released to farmers. In a hungry world one cannot neglect trying to raise productivity, but a project to breed high-yielding cultivars is not something to be undertaken in isolation.

11.5   THE VERTIFOLIA EFFECT AGAIN

The vertifolia effect was discussed in Chapter 8. It is the erosion of horizontal resistance when new cultivars are bred and selected from breeding material protected against disease by vertical resistance. It was concluded in that chapter that high-yielding ability and high horizontal resistance to disease are often incompatible. The present chapter gives one reason why.

11.6   LOW-SUGAR DISEASE PROCESSES

Maize stalk rot and tomato Alternaria blight are examples of what have been called low-sugar disease by Horsfall and Dimond (1957). They are associated with low-sugar susceptibility or high-sugar resistance.

The concept of low-sugar disease has been supported by work on *Helminthosporium* leaf spots of grasses. Lukens (1970) worked with the disease of Kentucky bluegrass (*Poa pratensis*) caused by *H. sorokinianum* and observed the natural occurrence of the disease in turf. Susceptible varieties had less sugar than resistant varieties; and when the sugar content was lowered by shade, susceptibility increased. Vidhyasekaran (1974) worked with the disease of finger millet (*Eleusine corocana*) caused by *H. tetramera*.

He reported results in terms of a disease index (0 = no infection; 4 = severe disease). In continuous light the content of total soluble sugars was 21.8 mg/g of fresh weight and the disease index 0.1. In continuous darkness the figure was 10.7 mg/g and the index 2.0. With alternating 12 h of light and 12 h of darkness the figure was 12.3 mg/g and the index 1.2. An additional 90 kg/hectare of nitrogen reduced sugar from 14.9 to 8.6 mg/g, and increased the index from 0.7 to 2.4. Spraying leaves with a 10% glucose solution increased the soluble sugar content in uninoculated leaves from 12.5 to 19.6 mg/g and reduced the disease index in the inoculated leaves from 2.2 to 0.6. Robinson and Hodges (1977) worked with *Poa pratensis* and *H. sorokinianum*. They reported results in lesion-type ratings (1 = flecks, 16 = lesions that coalesced between points of inoculation). They compared plants fertilized and unfertilized with extra nitrogen and compared different forms of nitrogen. None of the nitrogen treatments changed the number of lesions per unit area of leaf, but there was a change in lesion type. Treatment with ammonium sulfate decreased the total sugar content (in μmol/g of fresh weight) from 41.19 in the control plants to 31.40 and increased the lesion-type rating significantly from 27.4 to 32.2. With ammonium nitrate the corresponding figures were 35.01 for sugar content and 29.8 for lesion type. Gibbs and Wilcoxson (1972) also worked with *Poa pratensis* and *H. sorokinianum*. They claimed results contrary to those of Lukens (1970), but in point of fact a close reading shows their findings to be in line with those of the others who worked with *Helminthosporium* leaf spots.

Horsfall and Dimond's (1957) concept of low-sugar disease and high-sugar disease must be modified. Replace the word "disease" by the word "process" or words "disease process." Instead of writing of "low-sugar disease," write of "low-sugar disease processes." These are usually, perhaps always, necrotrophic. Instead of writing of "high-sugar disease," write of "high-sugar disease processes." These are usually but not always biotrophic. (The rotting of sweet grape berries by *Botrytis cinerea* is an example of necrotrophy amidst high sugar.) There are two reasons for writing of processes instead of diseases.

First, many pathogens such as *Phytophthora infestans* or *Xanthomonas malvacearum* are hemibiotrophs. They attack biotrophically at the start and then switch to necrotrophy. A single disease can involve two opposite processes and two opposite responses to sugar. To the experimenter this poses a problem. To the hemibiotroph it gives the chance of making the best of both worlds. Host plant cells can first be attacked biotrophically when they are sugary and then necrotrophically after biotrophy has depleted them of sugar.

Second, there is a stage in the infection processes by fungi that is neither biotrophic nor necrotrophic. Biotrophy and necrotrophy imply feeding on

the host cells. The germinating spore and young sporeling feed nonpara-
sitically on food reserves within the spore or saprophytically on exudates or
other surface material. Up to the time when the pathogen begins to feed
parasitically in the host, the sugar content is not relevant either to necrotrophy or biotrophy. As a test of the concept of a low-sugar disease process,
the number of lesions established per unit area of host surface after artificial
inoculation is either inept or irrelevant.

It was with the establishment of infection that Gibbs and Wilcoxson (1972)
were concerned with one exception, and that exception supports the concept
that the infection of *Poa pratensis* by *Helminthosporium* is a low-sugar
process. In low light the sugar content of the leaves was reduced. This did not
materially affect the number of lesions established after artificial inoculation;
but it did allow the established lesions to grow faster and larger, as the
concept predicts.

11.7   THEORIES ABOUT HIGH-SUGAR RESISTANCE

Basically there are three theories about high-sugar resistance. All of them
have defensible background support.

First, a high-sugar content depresses the enzymatic processes involved in
degrading cell walls and protoplasts. The inhibition of enzyme synthesis by
high levels of sugars has been studied in bacteria, yeasts, and pathogenic
fungi (Anderson and Wood, 1969; Lukens 1970; Magasanik, 1961; Paigen
and Williams, 1970; Patil and Dimond, 1968; Sands and Lukens, 1974; and
Vidhyasekaran, 1974). Most low-sugar disease processes, i.e., those associated with high-sugar resistance, are necrotrophic; and necrotrophy differs
from biotrophy largely in involving much tissue degradation. There is
therefore a defensible background in attributing high-sugar resistance to the
inhibition of enzymes concerned with degradation. On the other hand, it is
difficult to maintain the theory of high-sugar inhibition when one thinks of
how efficiently *B. cinerea*, for example, rots grape berries. Perhaps more
attention should be given to the oxidation and degradation of lignin.

Second, a high-sugar content supports a high content of antibiotic
substances in host cells. Benzoxazinone glucoside occurs in maize stalks,
and decreases of this compound have been correlated with decreases in sugar
level and with cell death (Fajemisin and Hooker, 1974; Wysong and Hooker,
1966). Insofar as the antibiotic substances interfere with enzyme activity, this
has the same defensible background as the previous theory.

Third, a high-sugar content staves off senescence; senescence weakens the
cells and makes them susceptible. The background is defensible. There is
unquestionably an ontogenic effect. Except for epidemic disease that starts

from very low levels of initial inoculum, a sink-induced loss of resistance is causally associated with disease that strikes late, while the sink is developing. Many will, however, question the use of the word "senescence." To them it seems inappropriate to describe a maize plant as senescent when its leaves are fully green and it is beginning to develop those kernels on which the livelihood of the farmer depends.

11.8   DISCUSSION

High potential yield is incompatible with high resistance. That is what sink-induced loss of resistance is about. With no crop have plant breeders been more busy than with maize. Millions of dollars are spent yearly on maize breeding by public institutions and seedsmen. Yet with stalk rots the incompatibility remains; the evidence of Dodd (1980b) and Koehler (1960) in Section 11.2 shows this. The highest yield in farmers' fields and experimental plots is a compromise. The highest potential yield of plants free from stalk rot must be balanced against the risk of stalk rotting that high yield brings. The risk can be mitigated by cultural practices, balanced fertilizer, and more resistant genotypes, but it cannot be eliminated.

The resistance involved in sink-induced loss is horizontal. There is no evidence for sink-induced loss of vertical resistance. Sink-induced loss is associated with high-sugar resistance. Resistance to necrotrophic processes is almost certainly high-sugar resistance, and is horizontal. The plant breeder does not necessarily have a choice of resistance. Resistance to stalk rot in maize is probably only horizontal, because rotting is a necrotrophic process. The chance of finding vertical resistance to maize stalk rot seems remote.

To summarize, sink-induced loss of resistance is probably loss of horizontal resistance against diseases prevalent late in the season when the sink develops. A catalog of diseases in which loss of resistance is sink-induced would be useful. The test for sink induction is simple: Remove the sink and see what happens.

12

High-Sugar Disease Processes and Biotrophy

12.1 INTRODUCTION

That different diseases respond differently to sugar is knowledge that goes back to Yarwood (1934). He excised clover leaflets and floated them on sugar solutions. Floating on a 10% sucrose solution as compared with water or a 2% solution made leaflets more susceptible to rust caused by *Uromyces fallens* and mildew caused by *Erysiphe polygoni* but less susceptible to target spot caused by *Stemphylium sarcinaeforme* and anthracnose caused by *Colletotrichum trifolii*. Horsfall and Dimond (1957) surveyed the evidence and classified disease into high-sugar disease and low-sugar disease. To account for hemibiotrophs which are probably favored sequentially by high sugar in their biotrophic phase and by low sugar in their necrotrophic phase, it was suggested in Chapter 11 that it would be better to replace the word "disease" by "disease process."

Biotrophy and necrotrophy inevitably enter the discussion. The most notable high-sugar diseases, the rusts and powdery mildews, are biotrophic. The low-sugar disease discussed at length in the previous chapter, maize stalk rot, is necrotrophic. Powdery mildews introduce a special complication. Resistance is reversed. As sugar content increases, high-sugar susceptibility is replaced at still higher sugar contents by high-sugar resistance. Osmotic phenomena intervene. Sugar, which up to a point is food or an aid to feeding, becomes beyond that point an osmotic toxin, raising osmotic pressures to levels powdery mildews cannot tolerate. This was worked out by Trelease and Trelease (1929) more than half a century ago, and their work, had it had the attention it merited, would have removed much confusion from the literature. As will be discussed in more detail later, Trelease and Trelease found that wheat powdery mildew is favored by increased sugar content up to 0.8 M and is then inhibited by still higher contents. The optimal content (0.8 M) holds, regardless of whether the solute is sucrose, glucose, or glycerol. It is not clear whether the rusts and other biotrophs are as sensitive to osmotic pressure as the powdery mildews. Be this as it may, the intervention of osmotic pressure into what would otherwise have been a comparatively simple account of the relation of sugar to biotrophy has necessitated the inclusion of a special section in this chapter.

12.2   HIGH-SUGAR SUSCEPTIBILITY

Fromme (1913) showed that *Puccinia coronata* needed light for its development on oats. Development was retarded in proportion to the length of time the plants were kept in the dark, and this held, irrespective of whether light was excluded at the time of inoculation or after the pathogen was established. Reed (1914) showed that *Erysiphe graminis* failed to infect cells of wheat or barley that did not contain chlorophyll. Mains (1917) showed that it was not the chlorophyll as such that mattered; working with *P. sorghi* on maize and *P. coronata* on oats, he showed that seedlings or pieces of leaves kept in the dark became susceptible if they were supplied with sugar.

Trelease and Trelease (1929) worked with *E. graminis* on wheat leaves floated on sugar solutions in the dark. Melezitose, sucrose, fructose, and glucose all made the leaves susceptible to infection. Three of their findings need special mention.

First, increasing concentration of sugar increased the development of *E. graminis* up to an optimum. This optimum was 27.37 g of sucrose/100 ml of solution, 14.41 g of glucose/100 ml, and 7.36 g of glycerol/100 ml. These concentrations correspond approximately to a 0.8 M solution or 19.3 atm. Osmotic properties are evidently involved; and Trelease and Trelease suggested that at concentrations above the optimum there might be injury through plasmolysis, a low water-supplying power of the solution, or a high vapor pressure deficit of the atmosphere above the solution.

Second, the increase of mildew developed below the optimum is more closely related to the carbon concentration of the solution than to its osmotic pressure. Thus, for concentrations below the optimum the capacity of the solutions to promote mildew development is nearly the same for the different sugars at equal carbon concentrations.

Third, all their data indicate that (below the optimum) a large surplus of carbohydrate in the host leaf is needed for the abundant development of mildew. For abundant development, the supply of carbohydrate must be considerably more than that needed for the retention of life by the leaf. Mildew is abundant on leaves on a 0.1 M solution of glucose, yet susceptibility continues to increase to an optimum on a solution of 0.8 M sucrose, with 16 times as much carbon. This concentration of sucrose at the optimum is much higher than that usually needed to overcome substrate limitation of respiration, and it would seem that an increased supply of sugar increases susceptibility in ways other than by supplying substrates for respiration. That is, it seems that sugars are not only food but that they also promote mildew development in some other way.

In addition to this direct evidence for the association of rust and mildew development with sugar content, there is indirect evidence for an association with carbohydrates. "Green islands" surround sites of rust and mildew infection. The literature of chlorophyll regulated by the pathogen is large and goes back to Cornu (1881). In "green islands" photosynthesis seems to proceed normally, and in effect the pathogen ensures its own supply of photosynthate.

In plants infected with rusts and mildews host nutrients accumulate at sites of infection. This occurs either by the blocking of photosynthate transport away from infected leaves or by the transport of photosynthate from healthy to infected leaves or parts of leaves. Auxins, cytokinins, and gibberellins are hormones that may be involved in this. The body of literature is large and has often been reviewed.

12.3 THE SUGAR EFFECT

Sugar is food. The more sugar, the more food. Biotrophs need more sugars in the host cell than necrotrophs do. They must keep the host cell alive and can therefore extract only that amount of sugar or other nutrients surplus to the host cell's own minimum needs. Necrotrophs on the other hand use the host cell itself as food.

High sugar has disabling effects, as inhibitors of the synthesis of enzymes of degradation and as substrates for the synthesis of tannins and other inhibitors of enzyme action. But these disabilities hardly apply to biotrophs because biotrophy, unlike necrotrophy, involves the minimum of cell degradation.

Sugars promote the polymerization of protein and therefore, on the protein-for-protein hypothesis, the protein food supply of the biotroph. Here we come to the observation of Trelease and Trelease (1929). To recapitulate, they found that the concentration of sucrose at the optimum for mildew development is much higher than that usually needed to overcome substrate limitation of respiration. Does this extra sugar increase mildew development as extra food or by enhancing protein polymerization? Trelease and Trelease's results are consistent with the possibility of enhanced protein polymerization by enhanced sugar content. Sugar enhances protein polymerization by altering the structure of the aqueous solvent; sugar makes hydrogen bonds with water. The difference in O:C ratio between glucose and sucrose is too small to be distinguished in the results of Trelease and Trelease. Their findings that mildew development is proportional to the C content of the solution on which the leaves floated could also be interpreted to mean that mildew development was proportional to the O content of the solute. A modification of Trelease and Trelease's experiment might answer the question: Float wheat leaves on a basic solution of 0.1 M glucose, with varying concentrations (less than 0.7 M) of a sugar or a polyol that is not toxic to host or mildew and cannot be metabolized as food by host or mildew.

12.4   REVERSAL OF RESISTANCE TO POWDERY MILDEW

Trelease and Trelease (1929) found that when they floated wheat leaves on a sugar solution, mildew development increased with increasing sugar concentration up to an optimum of 0.8 M. Thereafter, there was a reversal, and increased sugar concentration above the optimum made the leaves resistant. Sugar, which in low concentrations is a food, becomes an osmotic toxin for wheat mildew at high concentrations. It so happens that in many host plants of mildew the sugar concentration fluctuates near the optimum; the plants or parts of plants fluctuate from susceptibility to resistance as the sugar concentration fluctuates from below to above the optimum. The optimum depends on the mildew species and host plant, and presumably on the environment.

Weinhold and English (1964) studied mildew of peaches caused by *Sphaerotheca pannosa*. Young leaves are susceptible, old leaves resistant. Weinhold and English found that the apical leaf, which is susceptible, had sap with an osmotic pressure of 16.2 atm. The leaf of the first node was also susceptible; the sap had an osmotic pressure of 16.8 atm. The leaves at the second and third nodes were resistant, and the osmotic pressures were 20.7 and 22.0 atm, respectively. Mature, resistant leaves could be made susceptible by keeping them in the dark until the sugar concentration fell.

Delp (1954), studying powdery mildew of grapes caused by *Uncinula necator*, found that berries of several varieties became resistant after they matured to 8% sugar or more.

Schnathorst (1959), studying *E. chicoracearum* in lettuce, concluded that the osmotic values of leaves of field-resistant plants or of resistant leaves of field-susceptible plants were great enough to inhibit water uptake and growth of the fungus. Resistance due to the high osmotic pressure of cell sap explains the following: resistance of young seedlings in the field, gradients of mildew infection on plants in the field, increased susceptibility of lettuce infected with lettuce mosaic virus, resistance of lettuce grown in cooler areas of the Salinas Valley in California, increased susceptibility of shaded plants, changed resistance by mineral nutrition, change from resistance to susceptibility in detached lettuce leaves cultured in distilled water, failure of mildew to infect guard cells, and field resistance of some cultivars.

Probably in no other group of diseases are the ontogenic effects on resistance so varied as in the powdery mildews. The conflicting effects of sugar as food and as an osmotic toxin explain why.

13

Epidemiological Effects of Vertical Resistance

13.1 RESISTANCE IN RELATION TO AN INCREASE IN THE POPULATION OF THE PATHOGEN

This chapter deals with the effect of vertical resistance on the progress of disease in fields. Vertical resistance manifests itself as a delay in the start of the epidemic, without necessarily changing the infection rate after the start.

For simplicity, the discussion is built mostly around potato blight caused by *Phytophthora infestans*. The *R* genes are pseudomonogenes, i.e., the resistance to avirulent races and susceptibility to virulent races are complete.

An epidemic is an increase in the population of the pathogen. *P. infestans* survives the winter in diseased potato tubers. The amount that survives is usually relatively very small; but during the summer, if the weather is favorable for disease and crops are not protected by fungicides, the fungus can increase a billionfold in an epidemic that destroys the foliage of susceptible potato varieties. The more susceptible the variety, the earlier is the destruction of the foliage.

Resistance in the host curbs the epidemic. It reduces the level reached by the population of the pathogen. The purpose of using resistance is to bring about this reduction, and by so doing, to prevent substantial damage to the host.

The study of resistance, whether in its theoretical or practical aspects, necessarily includes the study of the effect of resistance on the population dynamics of the pathogen. This is part of epidemiology, and the ordinary concepts of epidemiology can be applied. To make what may be unfamiliar concepts easier to understand, the discussion in the first part of this chapter, as far as the end of Section 13.6, is in broad outline without any mathematics. Thereafter, the discussion is reinforced with some elementary calculations, and proofs and demonstrations are given for the bare statements made earlier.

13.2   THE EFFECT OF VERTICAL RESISTANCE: THE GENERAL RULE

The general rule for pathogens such as *P. infestans* is that vertical resistance in the host delays the start of an epidemic. This rule must be qualified in parts but is general enough to serve as the basis for this chapter. The qualifications will be mentioned later.

The action of vertical resistance is to reduce the effective amount of initial inoculum from which the epidemic starts and thereby to delay the observed start. To introduce the topic, let us begin with a hypothetical example and return to experimental evidence later.

Suppose there are two potato fields, side by side. In the one field, the variety has no *R* gene; it lacks vertical resistance to blight. In the other, the variety has the gene *R1*; it has vertical resistance to many races. Blight (it is supposed) is rare early in the season—relatively little of the fungus survives the winter—and both fields are healthy at the start. Later they are reached by light showers of spores originating from fields that had become infected earlier. Of the spores in the showers suppose that 99% belong to races that cannot infect the *R1* type: races (0), (2), (3), (4), (2, 3), ... These can infect only one of the fields: the field without vertical resistance. The other 1% of the spores (it is supposed) belong to races that can infect an *R1* type: races (1), (1,2), (1,3), (1,4), (1,2,3), ... These can infect both fields; to them the field with the gene *R1* is just as susceptible as the field without it. The result of the spore showers is that the field without the *R* gene starts with 100 times as much effective inoculum as the field with the gene. The initial number of lesions (per plant or square yard or acre, whatever unit one chooses) is 100 times as great in the field without the *R* gene as in the field with it; this statement involves the relation between the number of spores and the number of lesions and will be substantiated later in Section 13.8. Vertical resistance given by the *R* gene has reduced the initial inoculum to $1/100$ of what it otherwise would have been. From the initial lesions the fungus starts to increase: The epidemic starts. It proceeds as fast in the field with the *R* gene as in the field without it, but the amount of inoculum in the field with the R gene is only 1/100 of that in the other field. Because of the smaller amount of initial inoculum, the epidemic in the field with the R gene lags behind by the period needed for the inoculum to catch up, i.e., by the period needed for disease to increase 100 times. The reduction of effective initial inoculum brought about by the R gene shows itself as a delay in the epidemic.

Figure 13.1 illustrates this with hypothetical data. The showers of spores are supposed to have arrived in the second half of July. The amount of disease they caused is supposed to have been small, affecting, say, only 0.1% of the foliage of the susceptible variety, the variety without the R gene, and 0.001% of the foliage of the resistant variety, the variety with the R gene. (These amounts correspond roughly to one lesion per plant of the susceptible variety and one lesion per 100 plants of the resistant variety, and are realistic enough.) These levels are too low to show in the graph at the start; but disease is supposed to reach recordable levels later, at the beginning of August in the susceptible variety, and 10 days later in the resistant variety. That is, it is supposed that it took the disease 10 days to increase one hundredfold, so that in the resistant field, the disease took this period to make up the initial difference and become recordable. (A hundredfold increase of disease in 10 days represents a fast, but not unrealistically fast, rate of increase.) Thereafter, the two curves in Fig. 13.1 are shown to have the same form, the curve for the resistant field being at every level of disease 10 days behind the other. For example, Fig. 13.1 shows 50% disease as having been reached on August 13 in the susceptible field and on August 23 in the resistant field.

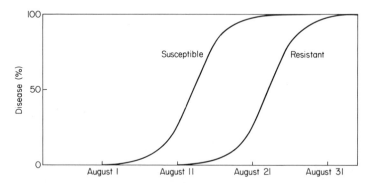

Fig. 13.1.   A representation of the effect of vertical resistance to potato blight conferred by an R gene. The curves show the progress of blight in two varieties. The susceptible variety has no R gene; all blight races infect it. It is supposed that only 1% of the spores reaching the foliage are of races that can infect the resistant variety. The infection rate is so chosen that blight increases 100-fold in 10 days during the early stages of the epidemic. Resistance therefore delays the blight epidemic by 10 days.

The assumption here, made for simplicity, is that the infection rate stayed constant.

After the initial infection has occurred, the rate of increase of disease is not reduced by the presence of R genes. The infection rate is as fast in the vertically resistant as in the susceptible variety. What this amounts to is that race (1), for example, can attack a variety with the gene $R1$ as easily as race (0) can attack a variety without an R gene; the spores germinate and penetrate in the same way, the mycelium spreads through the tissues in the same way, and spores are produced in the same way and in the same numbers. The whole infection process seems to be the same. An observer, even the most expert observer, shown a field of blighted potatoes could not decide just by examining the lesions whether the potato variety he was observing had R genes or not; nor would repeated visits to the field at intervals in order to determine how fast the disease was increasing help him to decide.

13.3 THE EFFECT OF VERTICAL RESISTANCE: SOME ILLUSTRATIVE DATA

The delay in an epidemic that vertical resistance brings about is illustrated by the data Kirste (1958) obtained at Celle, West Germany, in 1955 when blight was severe. These data are reproduced in Table 13.1.

Consider the first column of dates, which gives the dates when blight was first recorded. From these dates one can estimate the delay of the start of the epidemic that the R genes brought about. There were nine medium-early (ME) varieties. On the three varieties without R genes (Bona, Concordia, and Heideniere) blight was first recorded on July 25. On five of the varieties with the gene $R1$ the date was July 30, and on one of them, July 25. This gives an average of July 29. Judged by this average, the vertical resistance conferred by the gene $R1$ delayed the appearance of blight by 4 days, from July 25 to July 29. If one considers all the varieties without R genes and all with the gene $R1$ in Table 13.1 the average delay was 5 days.

Calculations in Section 13.9 show that this observed delay of 4 or 5 days is within the limits of what one would expect from the known relative abundance of race (1) and other races able to match the gene $R1$.

Another comparison is between the variety Vertifolia and the six late-maturing varieties (Ackersegen, etc.) without R genes. Vertifolia has vertical resistance conferred by the genes $R3$ and $R4$; and this delayed the start of the epidemic by 15 days, from August 7, the average date on which blight appeared on the six varieties, to August 22, the date on which it appeared on Vertifolia. This observed delay is also within the limits of what one would

TABLE 13.1

Progress of an Epidemic of *Phytophthora infestans* on Various Potato Varieties[a]

Variety	Maturity[b]	Major genes[c]	Date of reaching blight rating[d]			
			1	2	3	4
Bona	ME	0	July 25	—	July 30	Aug. 16
Concordia	ME	0	July 25	July 30	Aug. 13	Aug. 25
Heideniere	ME	0	July 25	—	July 30	Aug. 13
Forelle	ME	R1	July 30	Aug. 7	Aug. 13	Aug. 25[e]
Lori	ME	R1	July 30	Aug. 7	Aug. 10	Aug. 19
Augusta	ME	R1	July 30	Aug. 7	Aug. 13	Aug. 19
Cornelia	ME	R1	July 30	Aug. 7	Aug. 10	Aug. 22
Luna	ME	R1	July 30	Aug. 7	Aug. 13	Aug. 19
Suevia	ME	R1	July 25	July 30	Aug. 13	Aug. 19
Heida	ML	0	July 30	Aug. 7	Aug. 13	Aug. 28
Voran	ML	0	Aug. 7	Aug. 13	Aug. 28	—
Maritta	ML	R1	Aug. 7	Aug. 16	Aug. 22	Aug. 28
Margot	ML	R1	Aug. 7	Aug. 13	Aug. 16	Aug. 22
Benedikta	ML	R1	Aug. 7	Aug. 13	Aug. 22	—
Oda	ML	R1	Aug. 3	Aug. 13	Aug. 19	—
Apta	ML	R1	Aug. 3	Aug. 13	Aug. 22	Aug. 25
Nova	ML	R1	Aug. 10	Aug. 16	Aug. 25	Aug. 28
Urtica	ML	R1	Aug. 7	Aug. 13	Aug. 25	—
Virginia	ML	R1R4	Aug. 7	Aug. 10	Aug. 19	Sept. 8
Lerche	L	0	Aug. 3	Aug. 7	Aug. 25	—
Carmen	L	0	Aug. 7	Aug. 25	Sept. 4	—
Heimkehr	L	0	Aug. 3	Aug. 16	Aug. 28	—
Ackersegen	L	0	Aug. 3	Aug. 13	Aug. 28	—
Capella	L	0	Aug. 13	Aug. 16	Sept. 20[f]	—
Ronda	L	0	Aug. 10	Aug. 16	Sept. 1	Sept. 4
Ancilla	L	R1	Aug. 13	Aug. 16	Sept. 1	—
Monika	L	R1	Aug. 13	Aug. 19	Sept. 1	—
Adelheit	L	R1	Aug. 13	Aug. 22	Aug. 28	—
Herkula	L	R1	Aug. 13	Aug. 19	Sept. 4	—
Panther	L	R1	Aug. 13	Aug. 19	Aug. 25	—
Vertifolia	L	R3R4	Aug. 22	—	Aug. 28	Sept. 1

[a] From Vanderplank (1963, p. 186): adapted from a table of Kirste (1958). For convenience this table incorporates Table 8.2.

[b] Here, ME stands for medium early, ML medium late, and L late in maturity. Classification of Schick *et al.* (1958a).

[c] No major gene present = 0.

[d] Ratings: 1 indicates blight first recorded, only occasional lesions found; 2 indicates mild attack, lesions on about one leaf per plant; 3 indicates medium infection, several leaves per plant attacked; 4 indicates all leaves attacked, but the plants still generally green.

[e] Forelle was still at stage 3 when last examined on August 25.

[f] Capella was still at stage 2 when last examined on September 20.

expect from the known relative abundance of races able to match the genes
R3 and *R4* (see Section 13.9).

The much greater delay of the epidemic of blight in the variety Vertifolia
than in the varieties with the gene *R1*, i.e., the much greater vertical resistance
of the variety Vertifolia, resulted from the relative scarcity of races that
could attack Vertifolia with the genes *R3* and *R4* compared with races that
could attack varieties with the gene *R1*. This does not mean that the genes
R3 and *R4* are better than *R1*. Essentially, the reason is probably historical.
Breeders used the gene *R1* before *R3* and *R4*, and more widely; consequently,
races able to attack *R1* types are more common than those able to attack
R3R4 types.

The other three columns of dates are relevant to the present topic because
they prove the point that the presence of an *R* gene does not slow down an
epidemic once it has started. The infection rate can be judged by the time
taken for blight to increase from rating 1 to rating 2, from rating 2 to 3, or
from rating 3 to 4 in Table 13.1. For example, 9 varieties without *R* genes
and 19 with *R* genes are shown with both ratings 1 and 2 in Table 13.1.
Blight took an average of 8 days to increase from rating 1 to 2 in the varieties
without *R* genes, but only 7 days in the varieties with *R* genes. It increased
slightly faster in the varieties with *R* genes. Similarly, blight took an average
of 16 days to increase from rating 2 to 3 in the varieties without *R* genes,
but only 8 days in the varieties with *R* genes. Again, blight increased faster
in the varieties with *R* genes. Evidently, there is a vertifolia effect (see Sections 8.1, 8.2, and 11.5). It is wished only to establish that the whole benefit
from *R* genes lies in the way they delay the start of an epidemic. There is no
known evidence of any other benefit conferred by *R* genes.

To revert to the estimate given some paragraphs back, vertical resistance
given by the gene *R1* delayed the epidemic in Kirste's plots in West Germany
in 1955 by an average of only 5 days. This is relatively little. It reflects the
fact that by 1955 varieties with the gene *R1* had become widely grown in
West Germany. Inoculum that survived the winter in tubers of these varieties
was of races virulent on *R1* types; and inoculum of this sort was abundant
in proportion to the abundance of the varieties. In the Netherlands, however,
there were then no commercial varieties of any importance with the gene
R1, and the fields were almost exclusively of varieties without *R* genes.
Virulent races were consequently relatively rare. Therefore, the vertical
resistance given by *R* genes was proportionately more effective and delayed
the start of epidemics by a much longer period. Some figures are given by
Toxopeus (1956). In varieties without *R* genes, blight was prevalent by the
middle of July; it can be assumed that the epidemics started to be noticed
early in July. In varieties with the gene *R1*, the epidemics started to be noticed
early in August; in varieties with the gene *R3*, in the second half of August;

and in varieties with both the genes *R1* and *R3*, at the end of September. Vertical resistance given by the gene *R1* delayed the start of the epidemic by about 1 month; by the gene *R3*, about 1.5 months; and by the genes *R1* and *R3* together, 2.5 months or more. These figures should be enough to prove that the beneficial effect of resistance, even though it is limited to delaying the start of an epidemic, can be substantial. The condition for a large beneficial effect is that the relevant races should be rare (see also Section 13.11). The condition was met. Races able to attack varieties with both the genes *R1* and *R3* were known in the Netherlands in 1956 but were very rare. [The known races able to do this were (1,3), (1,3,4), and (1,2,3,4). They were sufficiently rare to support Toxopeus' suggestion that they may have arisen during the season by mutation from other races. This suggestion need not be probed here because it does not affect the discussion. The concern is with initial inoculum, wherever it comes from.]

13.4   VERTICAL RESISTANCE: THE PRICE OF VARIETAL POPULARITY

To represent the effect of vertical resistance in Fig. 13.1 it was supposed that the variety with the *R* gene could be attacked by only 1% of the spores reaching the foliage, i.e., only 1% of the spores were of races to which the variety was susceptible. The infection rate was chosen so that this meant that resistance delayed the epidemic by 10 days. If 10% instead of 1% of the spores had been of races able to attack the variety with the *R* gene, the delay (with the same infection rate) would have been 5 days. If 50% had been of races able to attack, the delay would have been 1.5 days. If all the spores had been of races able to attack the variety with the *R* gene, there would have been no delay; the variety with the *R* gene would have been fully susceptible; the *R* gene would no longer have conferred resistance.

As virulent races become more abundant, vertical resistance wanes. The main reason for virulent races becoming abundant is the extensive planting of resistant varieties. As farmers grow more of a resistant variety, they may by doing so destroy much of the resistance and hence the reason for growing the variety. The increasing popularity of a variety among farmers can be a self-destructive process when the popularity depends on vertical resistance.

How the potato variety Kennebec with the gene *R1* came to lose its resistance to *P. infestans* as it became popular has been discussed in Section 7.6.3.

The way in which virulent races become prevalent when farmers grow vertically resistant varieties on a large scale has been observed for *Puccinia graminis tritici*. In Australia the wheat varieties Eureka with the gene *Sr6*,

Gabo with the gene *Sr11*, and Mengavi with the gene *SrTt1* all succumbed to stem rust as they became popular and virulent races prevalent (see Section 5.2 and Figs. 5.1, 5.2, and 5.3).

Varietal popularity, more than the mutability of the pathogen (see Chapter 6), is the cause of boom-and-bust cycles. Coupled with varietal popularity is the ability of the pathogen to survive on the variety or relevantly related host genotypes. Vertical resistance in the potato to *Phytophthora infestans* is quickly lost as the potato variety becomes popular because the fungus can overwinter in tubers used as seed or dumped near potato fields in the spring. The fungus need not leave the host genotype. Vertical resistance to *Puccinia graminis tritici* on wheat in Australia has repeatedly been overcome because the fungus survives the off-season in uncultivated grasses and volunteer wheat plants.

13.5   VERTICAL RESISTANCE: THE ENHANCING EFFECT OF HORIZONTAL RESISTANCE

Abundance of appropriately virulent races is only one of the factors in vertical resistance. Another is the infection rate.

In the hypothetical example on which Fig. 13.1 is based, the resistant variety is taken to be susceptible to 1% of the spores—those belonging to races virulent on the variety—and at the infection rate considered, this delays the epidemic by 10 days. This period of 10 days is the time taken for disease to increase 100 times. But if the infection rate had been halved, the delay would have been 20 days—vertical resistance would have been twice as effective—because disease would have taken 20 days to increase 100 times at the halved rate. This proposition is discussed quantitatively in Section 13.9.

Fast infection rates decrease the efficiency of vertical resistance; slow rates increase it. Weather is one of the factors influencing the infection rate. Weather that favors disease increases the rate and decreases the efficiency of vertical resistance; weather unfavorable to disease increases the efficiency.

More important for this narrative is horizontal resistance. Horizontal resistance that reduces the infection rate thereby enhances vertical resistance.

Consider the four hypothetical varieties represented by Fig. 13.2. Variety A, represented by curve A, has little horizontal resistance and no vertical resistance. Variety B, represented by curve B, has the same small amount of horizontal resistance as variety A, but has vertical resistance. This vertical resistance is enough to delay the epidemic in variety B by 10 days. (Curves A and B are from the two curves in Fig. 13.1.) Variety C, represented by

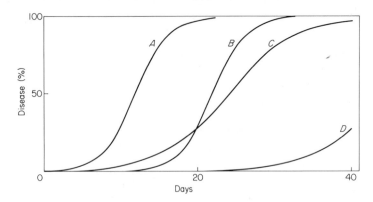

Fig. 13.2.   A representation of the effects of vertical and horizontal resistance, separately and combined on disease progress curves. Variety A, represented by curve A, has little horizontal and no vertical resistance. Variety B has the same small horizontal resistance as variety A, but it has vertical resistance. Variety C resembles variety A in having no vertical resistance, but it has considerably more horizontal resistance. Variety D combines the vertical resistance of variety B with the horizontal resistance of variety C.

curve C, resembles variety A in having no vertical resistance, but it has considerably more horizontal resistance. This extra horizontal resistance is enough to halve the infection rate; disease in variety C takes twice as long as in variety A to increase from one level to another (e.g., disease in variety C takes 9.6 days to increase from 10 to 50%, whereas in variety A it takes only 4.8 days). Variety D, represented by curve D, has both vertical and horizontal resistance; it has the same vertical resistance as variety B and the same horizontal resistance as variety C. Curve D (as far as it goes) has the same slope as curve C because the horizontal resistance is the same. But whereas curve B is only 10 days behind curve A, curve D is 20 days behind curve C because horizontal resistance has halved the infection rate and doubled the time needed for disease to make up the loss of initial infection caused by vertical resistance.

The horizontal resistance of variety D has greatly enhanced the variety's vertical resistance. Although neither the vertical nor the horizontal resistance is very great, as shown by curve B for vertical resistance alone and curve C for horizontal resistance alone, the combined resistances are very effective. Variety D becomes diseased only late in the season, too late for much harm to be done.

The good results obtained by combining vertical and horizontal resistance are important to know. The literature of resistance to disease contains much argument about whether vertical or horizontal resistance should be used and discusses the two kinds of resistance as alternatives. It seldom recognizes that the two kinds are best used in combination, that the one helps the other.

13.6   GENERALIZED DISEASE PROGRESS CURVES
FOR THE STUDY OF THE EFFECTS OF RESISTANCE

The curves in Fig. 13.2 represent reasonably realistically the progress of blight in fields of unsprayed potatoes. But the curves in Fig. 13.2 do not resemble the curves for the progress of, say, stem rust in wheat. If one takes the end of the graph to represent the end of the growing season, curve A in Fig. 13.2 represents stem rust severely damaging wheat before it even has had time to flower. This can, of course, happen; but is unlikely in commercial fields because wheat could not be grown commercially in climates where this is likely to happen.

Figure 13.3 is more realistic for wheat stem rust than is Fig. 13.2. Curve A in Fig. 13.3, for stem rust in a susceptible variety, represents severe infection. The variety ends the season with 95% disease, which means the yield is poor and the grain shriveled; but this is not unusual with susceptible varieties in many wheat-growing regions. Varieties B and C, with a modicum of vertical and horizontal resistance, respectively, end the season with 20–30% disease. Curve D of Fig. 13.2 is absent from Fig. 13.3. The combination of horizontal and vertical resistance in variety D practically eliminates rust, i.e., it delays the epidemic for so long that the crop can ripen without

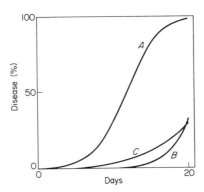

Fig. 13.3.   A representation of the effects of vertical and horizontal resistance on disease progress curves. Curves A, B, and C have the same meaning as in Fig. 13.2; but Fig. 13.3 differs from Fig. 13.2 in that the epidemic is supposed to have started 20 days later in the season. The end of the curves represents the end of the season when the crop is mature.

being substantially rusted. Rust, if present, occurs only in trace amounts too small to be recorded in the graph.

Examined superficially, Figs. 13.2 and 13.3 may seem different, but Fig. 13.3 is really just the first half of Fig. 13.2. Figure 13.3 represents an epidemic in which disease in susceptible varieties starts to become evident 40 days before the end of the season (or 40 days before the crop ripens if one prefers to think of it in this way). Figure 13.3 represents an epidemic in which disease in susceptible varieties becomes evident 20 days before the end of the season. Otherwise the two figures are identical. Which figure is more appropriate depends on the season's weather (epidemics starting earlier in "disease weather"), the amount of inoculum available (epidemics starting earlier when much inoculum is present), and any other factor that affects the timing of epidemics. The earlier an epidemic starts the more likely it is to be represented by Fig. 13.2.

As an exercise take a card and cover the second half of Fig. 13.2. What remains uncovered is Fig. 13.3. Move the card further to the left, to cover everything after, say, day 10. What remains of the figure represents an epidemic starting in susceptible varieties as late as only 10 days before the end of the season. With the particular infection rate used to prepare the graph, 10 days is enough for disease to mount to about 30% by the end of the season in a susceptible variety (curve A); but the variety with vertical resistance (curve B) would end the season without substantial disease—the vertical resistance would have been adequate for practical purposes. Exposing only the first 10 days of Fig. 13.2 would leave curves realistically representing many wheat stem rust epidemics. Move the card to the right; the more it is moved, the earlier the epidemic is shown to start, and the more severe it is.

Figure 13.2 represents the effects of resistance on the course of explosive epidemics in a general manner. Details in figures of this type can be made to vary endlessly. The infection rate varies, and with it the slope of the curves. The amount of vertical resistance varies, and with it the distance between the curves. The shape of the curves is not always a good S. It is easy enough to adjust for details of these kinds. The essential point is that the progress of disease can be shown by curves, and resistance can be evaluated by the way in which it shifts these curves or changes their slope.

Disease progress curves are the simplest, and perhaps the best, way of showing how resistance in the host can be interpreted in terms of disease in the field; and for practical purposes this is the interpretation one needs. To say that a potato variety has the gene RI (or any other gene or genes) is all very well, but it tells one little unless the effect of the gene on disease in the field can be determined. How to determine the effect is what concerns this chapter.

13.7   THE COMPOUND INTEREST EQUATION:
LOGARITHMIC INCREASE OF DISEASE

In previous sections of this chapter conclusions were reached without discussing their mathematical background. It was desired to put the principles plainly and simply, without distractions. However, the way resistance affects the progress of disease in populations of host plants—in plots, fields, plantations, etc.—is so important to the study of resistance that at least the elementary mathematical background must be discussed. The rest of the chapter is taken up with this.

The basis of the discussion is the compound interest equation. Diseases such as potato blight or the cereal rusts increase at "compound interest" during the early stages of an epidemic. The process differs from that in the accumulation of money mainly because the "interest" is deemed to be added to capital as soon as it is earned, whereas interest on money is usually added at intervals, commonly of one year.

The equation is

$$x = x_0 e^{rt}$$

in which x is the amount of accumulated money or disease at time t, i.e., the amount after it has accumulated for t days or years or whatever the units of time may be; x_0 is the initial amount of money or disease (i.e., the amount at zero time), r is the rate of interest or the infection rate, and e is 2.718. The equation is for interest added as soon as it is earned.

The theoretical foundation for this equation in epidemics is solid. Suppose that the pathogen is a fungus, and that the fungus in lesions releases spores that move away and germinate elsewhere in the field to start daughter lesions. If the fungus in 1000 parent lesions starts N daughter lesions, and the fungus in 2000 parent lesions in identical conditions starts $2N$ daughter lesions—if in any given conditions the number of daughter lesions is proportional to the number of parent lesions—the equation can be used. Much experimental evidence for this proportionality was analyzed in detail by Vanderplank (1975).

Money or disease or anything else that increases at a rate proportional to the amount already accumulated is said to increase logarithmically, or exponentially, or at "compound interest." It is not implied in this statement that the rate stays constant. Just as the rate of interest on money may change from month to month with a changing financial climate, so the infection rate may change from day to day, even from moment to moment, as the weather and other factors vary. That is, r is variable, and not a constant (although it is convenient and accurate to use average values).

Disease increases logarithmically, i.e., at compound interest, only if infections are independent: only if they do not interfere with one another. Therefore, disease increases logarithmically only when the percentage of disease is small. It can be expected that the daughter lesions of a parent lesion will not interfere with one another or with the daughter lesions of other parent lesions only if they are widely scattered, i.e., only if the percentage of disease is small. As the percentage of disease increases, lesions begin to crowd one another: they overlap or compete for space. When that happens, the number of daughter lesions can no longer be proportional to the number of parent lesions, and logarithmic increase ceases. A simple extreme example illustrates this. Suppose that in a given time in given climatic conditions, the amount of disease in a field doubles, from 1 to 2%. The amount of disease is still small, even after it has doubled. But suppose the amount were high. Disease could not double from, say, 50 to 100%, in the same time and in the same conditions as it increased from 1 to 2%, because most spores released from the parents lesions would fall on tissues already infected and be wasted from the pathogen's point of view. The increase from 50% would be proportionately less than the increase from 1 to 2%, which means that disease does not increase logarithmically at high levels.

For practical purposes one can set 5% disease as the limit below which increases can be considered logarithmic, and the compound interest equation applied with adequate accuracy. Below this limit the overlapping of lesions (or, for systemic diseases, the multiple infections of plants) can usually be ignored for practical purposes.

An objection can be raised: Lesions (or systemically infected plants) are not scattered uniformly over a field. They tend to occur in foci: areas of abnormally high concentration of disease. In these foci, overlapping of lesions (or other forms of mutual interference among lesions of systemically infected plants) is greater than the overall percentage of disease would suggest. Fortunately, it can be shown that this does not matter, that foci can be considered in the same way as lesions, and that as long as foci do not overlap considerably, the disease can be taken as increasing logarithmically. It is the average percentage of disease that matters in deciding what upper limit must be fixed for practical purposes. There is a qualification to this statement. Disease must be considered over an area large enough to include a large population of foci; the increase of a single focus is not logarithmic.

Two statements made in this section, that it is accurate to use average infection rates in the compound interest equation, and that this equation can be applied to populations of foci provided that the overall percentage of disease is small, were analyzed in detail by Vanderplank (1963).

A third problem, implied although not explicitly stated before in this section, is the period of latency. Newly infected tissue is not immediately infectious; it takes time before a newly started lesion itself forms spores, and this time is the period of latency. Nevertheless, disease in a population of plants increases at compound interest despite the period of latency; the length of the period affects the numerical value of the infection rate without affecting the form of increase, which remains logarithmic (Vanderplank, 1963).

In applying the compound interest equation it is convenient to use the proportion x of disease instead of the percentage (which is $100x$). Thus, $x = 0.01$ means 1% disease.

13.8   AN EQUATION FOR THE EFFECT OF VERTICAL RESISTANCE

The effect of vertical resistance in reducing initial inoculum is considered here. Suppose there are two fields of potatoes. The one field is of a variety without an R gene; the other is of a variety with the gene $R1$. Otherwise the fields are identical. Both fields are healthy at the start. Light showers of spores of *P. infestans* fall and infect them. Let the proportion of foliage infected by spores be x_0 in the field without the R gene and x_{0v} in the field with the R gene. The subscript v is for vertical resistance. The proportion x_0 is greater than x_{0v} because all races can infect the variety without the R gene, but only races (1), (1,4), etc. can infect the variety with the gene $R1$: the vertical resistant variety. From the initial inoculum, x_0 in the susceptible variety and x_{0v} in the resistant variety, disease starts to increase, and an epidemic starts to develop. The infection rate r is the same in both varieties. By hypothesis the conditions are identical for the two varieties, except that the one has the gene $R1$ and the other not; and the experimental evidence is convincing that races (1), (1,4), etc. increase just as fast in varieties with the gene $R1$ as races (0), (4), etc. do in varieties without an R gene (see Sections 13.2 and 13.3). The epidemic in the resistant variety lags behind that in the susceptible variety. It lags behind by the interval of time needed for disease to increase from the level x_{0v}, at which disease in the resistant variety started, to the level x_0, at which disease in the susceptible variety started. This interval of time Δt is given by the compound interest equation as

$$x_0 = x_{0v}e^{r\Delta t}$$

Hence

$$\Delta t = \frac{1}{r} \log_e \frac{x_0}{x_{0v}}$$

$$= \frac{2.30}{r} \log_{10} \frac{x_0}{x_{0v}} \tag{13.1}$$

For example, if $x_0/x_{0v} = 5$ and $r = 0.2$ per day then

$$\Delta t = \frac{2.3}{0.2} \log_{10} 5$$

$$= 8 \text{ days}$$

Figure 13.4 illustrates this numerical example. At all times the curve for the susceptible variety shows five times as much disease as the curve for the resistant variety. At every level of disease the curve for the resistant variety is $\Delta t = 8$ days behind the curve for the susceptible variety. An optical illusion may obscure this on the graph, but it can be verified by direct measurement.

The ratio x_0/x_{0v} is not only the ratio of initial inoculum in the susceptible variety to initial inoculum in the resistant variety, it is also the ratio of spores of all races to those of races virulent on the resistant variety. Thus, in the example just considered, $x_0/x_{0v} = 5$ means that 20% of the spores in the initial spore showers were of races able to attack the resistant variety. If the resistant variety had the gene $R1$, then 20% of the spores were of races (1), (1,4), and others able to attack the genotype $R1$; the other 80% were of races (0), (4), and others unable to attack the genotype.

This follows from the fact that at low concentrations of spores the number of lesions is proportional to the number of spores. For *P. infestans* in particular, Knutson and Eide (1961) showed that up to concentrations of spores

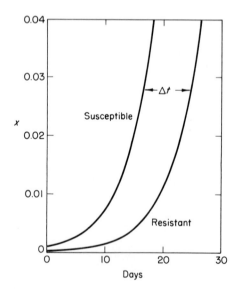

Fig. 13.4.   The delay of an epidemic brought about by vertical resistance. The ratio $x_0/x_{0v} = 5$; and $r = 0.2$ per day. The delay Δt is 8 days at all levels of disease. From Vanderplank (1963, p. 125).

that produced 250 lesions per potato plant the number of lesions was proportional to the number of spores. In other words, the spores acted independently of one another. The limit, 250 lesions per plant, amply covers all that is needed for the argument. For other pathogens the evidence is similar; it has been reviewed by Vanderplank (1963, pp. 89–93).

To end this section, here are details of the calculations underlying Figs. 131.1 and 13.2. The ratio $x_0/x_{0v} = 100$. For the two curves in Fig. 13.1 and curves A and B in Fig. 13.2, $r = 0.46$ per day; hence, $\Delta t = 10$ days according to Eq. (13.1). For curves C and D in Fig. 13.2, $r = 0.23$ per day; hence $\Delta t = 20$ days according to Eq. (13.1). The rate $r = 0.46$ per day is fast, but not abnormally fast; it is common in horizontally susceptible varieties in weather favorable for disease. This comment holds not only for potato blight but also wheat stem rust and many other diseases that cause explosive epidemics. The rate $r = 0.23$ per day is in a common range for varieties not very susceptible and weather not exceptionally favorable for disease.

13.9   AN ANALYSIS OF SOME EXPERIMENTAL DATA

Table 13.1 in Section 13.3 reproduced some data obtained by Kirste (1958). These data are for a potato blight epidemic in West Germany in 1955. The first column of dates shows when blight was first recorded, which one can take to be the start of the observed epidemic. An R gene in a variety delays the start of an epidemic in that variety. This can be checked simply by comparing the dates when blight was first observed; within any maturity class, varieties with an R gene developed blight later. The average delay brought about by the gene $R1$ was 5 days, and the delay brought about by the genes $R3$ and $R4$ in combination in the variety Vertifolia was 15 days. These estimates were given in Section 13.3. How do they tally with what one would expect from Eq. (13.1)?

To answer the preceding question one must first know the proportions of spores able to attack the genotype $R1$ and the proportion able to attack the genotype $R3R4$. Table 8.1 records observations made by Schick et al. (1958b) on the prevalence of races of P. infestans in Germany in 1956. (Unfortunately, appropriate data for 1955 are not available.) Schick et al. identified 209 isolates of P. infestans taken from plants of varieties without R genes. This detail is important, because the number of races that can be found on varieties with R genes is necessarily limited. The main races were (4), (1), and (1,4). The proportion of race (4) dropped as the season advanced. No reason was given. But the change is perhaps related to the fact that the most popular $R1$ varieties in Germany—varieties such as Maritta—were late in maturing.

This must affect the race population, even that reaching fields of varieties without R genes.

The other information needed to answer the question is the infection rate. One cannot determine this precisely from the definitions Kirste gave of the various blight ratings (these definitions are reproduced in Footnote d of Table 13.1), but the information he gave is enough to suggest that r was generally in the range from 0.23 to 0.46 per day.

With the above information Eq. (13.1) can be applied. Consider the figures for August 3 in Table 8.1. Races (1) and (1,4) were together 14.6% of the isolates. These can attack $R1$ varieties as well as varieties without R genes. The remaining 85.4% was race (4) which cannot attack $R1$ varieties. Therefore, $x_0/x_{0v} = 100/14.6$. With this ratio, Eq. (13.1) estimates Δt as 4 days if $r = 0.46$ per day, or as 8 days if $r = 0.23$ per day. That is, Eq. (13.1) estimates that the gene $R1$ would have delayed the start of the epidemic by 4 to 8 days according to the value of r chosen.

If one uses the data in Table 7.1 for August 29 instead of August 3, one must include race (1,2) which appeared in the interim. Races (1), (1,4), and (1,2) were together 30% of the isolates; thus, $x_0/x_{0v} = 100/30$, and $\Delta t = 3$ days if $r = 0.46$ per day, or 5 days if $r = 0.23$ per day.

To summarize, on the evidence available one expects from Eq. (13.1) that the benefit conferred by the gene $R1$ in West Germany would have been to delay the start of observed blight epidemics by 3 to 8 days according to the particular estimates selected for r and x_0/x_{0v}, and from Kirste's data the average observed delay was 5 days.

Other comparisons can be made. Vertifolia, a late variety, has the genes $R3$ and $R4$. Table 8.1 shows that Schick et al. (1958b) found 1 out of 209 isolates to be of a race—race (1,3,4)—able to attack it. With $x_0/x_{0v} = 209$, Eq. (13.1) estimates Δt as 12 days if $r = 0.46$ per day, or as 23 days if $r = 0.23$ per day. In another, larger survey in 1954 and 1955 (of $P.\ infestans$ in unspecified potato varieties), Schick et al. obtained 3 isolates of race (3,4) and 7 of race (1,3,4) out of a total of 979 isolates. With $x_0/x_{0v} = 979/10$, Eq. (13.1) estimates Δt as 10 and 20 days, respectively, for $r = 0.46$ and $r = 0.23$ per day. The delay observed by Kirste was 15 days, as already noted.

Thus, too, one can compare Virginia with the genes $R1$ and $R4$ with other medium-late varieties. The comparisons show no noteworthy discrepancies between theory and observation.

13.10   GRAPHICAL REPRESENTATION OF EQUATION (13.1)

Equation (13.1) relates Δt, the delay in the start of the observed epidemic, to r, the infection rate, and x_0/x_{0v}, the ratio of initial inoculum in the susceptible variety to that in the resistant variety (or, if this is more convenient, the

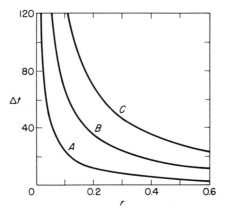

Fig. 13.5.   The relation between Δt, r, and x_0/x_{0v}. Curves A, B, and C are for ratios of 10; 1000; and 1,000,000, respectively. From Vanderplank (1963, p. 126).

ratio of the number of spores of all races to the number of spores of virulent races). Figure 13.5 shows the relation graphically.

The delay Δt is proportional to the logarithm of the ratio x_0/x_{0v} and is therefore not very sensitive to it. Curve A in Fig. 13.5 is for a ratio of 10, curve B for a ratio of 1000, and curve C for a ratio of 1,000,000. Yet the values of Δt for curves A, B, and C in Fig. 13.5 are only as 1:3:6 for every value of r.

The delay Δt is inversely proportional to r; and Fig. 13.5 shows how great is the delay, i.e., how beneficial is vertical resistance when r is small. Small values of r go with high horizontal resistance; and Fig. 13.5 shows in another way the activating effect of horizontal on vertical resistance, already considered in Section 13.5 and Fig. 13.2. Small values of r are also found in weather unfavorable to disease, and Fig. 13.5 brings out an essential weakness of vertical resistance against diseases such as potato blight. The effect of vertical resistance is strong in seasons of little disease when resistance is not very important, but weaker in seasons of much disease when resistance is most needed. It is a matter of history that all the great disappointments with vertical resistance have been in seasons of an unusually high infection rate.

13.11   VERTICAL RESISTANCE: THE QUANTITATIVE
EFFECT OF VARIETAL POPULARITY

In some (though not all) circumstances virulent races increase in abundance as the resistant varieties that they attack increase in popularity. Examples were given in Figs. 5.1, 5.2, and 5.3; as resistant wheat varieties increased in acreage in Australia, so did the races of *Puccinia graminis* that could attack them. The varieties left the breeder's nurseries as resistant, because races able to attack them were not known. Within a few years they were susceptible, because races able to attack them became abundant. This is the essence of the boom-and-bust cycle of variety production.

Figure 13.6 analyzes the process when the relative abundance of virulent races is proportional to the relative acreage of the resistant variety. (This state of affairs is illustrated by the wheat variety Mengavi, the subject of Fig. 5.3.) Figure 13.6 shows that at first when the variety has just been released and occupies only a negligible acreage of farmland, a condition designated as popularity zero in the graph, Δt is very great. Vertical resistance postpones the start of the observed epidemic for such a long time that it does not occur during the growing season of the crop; in more direct and concrete terms, there is no epidemic. The breeder is pleased, the farmers are pleased, and the disease seems at last to be under control. The following year farmers plant more of the variety—its popularity increases—and Δt decreases; Δt may still be large enough for no epidemic to be observed, i.e., the start of the observed epidemic may be delayed beyond the end of the season. But sooner or later, in some year or other, the popularity of the variety among farmers increases to the point at which Δt decreases to within the season's range. The epidemic is delayed, but it is observed to start before the end of the season. Finally, if the popularity reaches 100%, if farmers within the epidemic area plant no other variety, Δt drops to zero; vertical

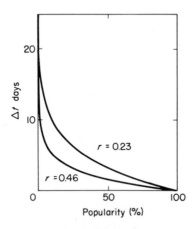

Fig. 13.6.   The relation between the benefit from vertical resistance and the popularity of a variety with this resistance. The benefit is measured as the number of days Δt by which an epidemic is delayed. Popularity (%) is the percentage of the acreage in a region planted to the resistant variety. The fast infection rate, $r = 0.46$ per day, is taken to represent a season climatically favorable for disease, and the slow rate, $r = 0.23$ per day, a season less favorable. From Vanderplank (1963, p. 182).

resistance fades away. Figure 13.6 shows the effects of two infection rates: 0.23 and 0.46 per day. Vertical resistance is lost faster at the higher infection rate.

The relations shown in Fig. 13.6 hold in their simplicity only when the percentage of virulent races follows proportionately the percentage of acres planted to resistant varieties, as happened with the wheat variety Mengavi, with gene *SrTt1* and stem rust. But the trend is general, even when it is not so simple. The popularity of a variety is the greatest destroyer of the variety's vertical resistance. A plant breeder is in a ludicrous position. He hopes his new variety will become popular among farmers. He hopes it will maintain its resistance. His hopes are likely to be mutually exclusive.

These remarks hold only for pathogens that can maintain themselves the year round on the variety, sown or self-sown. As pointed out at the end of Section 13.4, vertical resistance is not necessarily easily lost when the pathogen must exist on other varieties for part of the year or when it exists for part of the time in soil, crop residues, and the like.

13.12   THE EFFECT OF VERTICAL RESISTANCE AFTER THE LOGARITHMIC PHASE OF THE EPIDEMIC

In all calculations made so far we have used Eq. (13.1), which is derived from the compound interest equation. This holds only for the logarithmic phase of an epidemic, when the percentage of disease is small. How does Eq. (13.1) apply to an epidemic as a whole: to an epidemic that may range up to 100% disease?

Figure 13.7 shows the basic principle. The curves are the same as in Fig. 13.1, except that they have been broken near the base. Below the break the curves represent logarithmic increase of disease. Equation (13.1) can be applied to estimate Δt, the delay brought about by vertical resistance.

The upper part of the curves above the break represents an increase of disease which is increasingly different from logarithmic. An equation to estimate directly the delay brought about by resistance would be complicated. But for our purposes such an equation is unnecessary, provided that the disease progress curve has the same shape in the resistant as in the susceptible variety, whatever that shape might be. If the shape is the same, the delay Δt during the later stages of the epidemic—the Δt shown by the upper arrow—is the same as the delay at the beginning when the increase was logarithmic. If the shape stays the same, the position of the upper part of the curves is fixed by the position at the beginning, and Eq. (13.1) can be used generally.

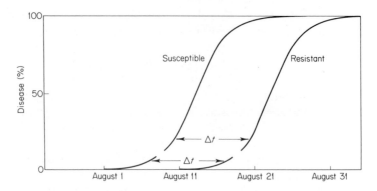

Fig. 13.7.   A representation of the relation between vertical resistance during the logarithmic phase of an epidemic and after this phase. The curves below the break are for the logarithmic increase of disease, and Δt can be estimated by Eq. (13.1). The curves above the break are not for logarithmic increase; but, provided the infection rate remains the same, Δt remains the same. From Vanderplank (1968, p. 49).

The reasoning holds only for epidemics that start, as most epidemics do, in the logarithmic phase of the increase of disease.

It is not implied that all epidemics, once started, proceed at the same rate. For our purposes it is enough to know what the effect of a change of rate would be. If the infection rate tends to slow down as the season advances, e.g., if the weather toward the end of the season tends to become less favorable for disease, or if internal factors reduce the susceptibility of the host as it grows older, the benefit of vertical resistance will be greater than that calculated from Eq. (13.1). That is one of the reasons why mature plant resistance is so valuable in crops protected by vertical resistance. Conversely, if the infection rate accelerates as the season advances, the benefit of vertical resistance will be less than that calculated by Eq. (13.1).

It is pure empiricism to consider disease progress curves to be S-shaped and the infection rate to be constant. It is, however, convenient and useful, and an S-shape is probably as good a way as any other to represent the progress of disease with time.

13.13   INDEPENDENCE OF INITIAL INOCULUM
AND THE LOGARITHMIC INFECTION RATE:
THE START OF AN EPIDEMIC

The logarithmic (exponential) infection rate is independent of the initial inoculum. That is the basis of Eq. (13.1). Independence is inherent in the definition of logarithmic increase.

If, as in the work of Gregory *et al.* (1981) and Rouse *et al.* (1981), at irrelevantly high proportions of disease, the infection rate and the initial inoculum are not independent, the increase is not logarithmic. The change from logarithmic increase to increase not logarithmic, i.e., from independence to dependence, comes when, with the increase of disease, lesions begin to interfere with each other. Multiple infection of lesions begins. This is why we calculate the delay in the start of an epidemic at as low levels of disease as possible. The delay at higher levels follows, as Fig. 13.7 shows.

The principle that vertical resistance genes, if they are fully effective against relevant avirulence in the pathogen, reduce initial inoculum and hence delay the start of an epidemic, is not in question. What is questionable is the accuracy of the estimate of the delay because of mutual interference between lesions.

It is probably unrealistic to fuss much about defining the start of an epidemic. There must inevitably be experimental errors in the estimated frequency of virulent races and in the estimated infection rates. In analyzing Kirste's data recorded in Table 13.1, the first column of dates was taken to show when the epidemic started. An even simpler criterion was used in Fig. 13.1. Epidemics were shown as starting when disease reached amounts recordable in the graph. This criterion is obvious and acceptable, despite the fact that the spore showers which were the initial inoculum were postulated as having arrived in the second half of July, and that these showers in turn must have come from fields infected even earlier.

13.14 APPENDIX: VERTICAL RESISTANCE THAT REDUCES THE INFECTION RATE

The opening paragraph of this chapter states that vertical resistance manifests itself as a delay in the start of an epidemic, without necessarily changing the infection rate. This is the general rule for pseudomonogenic resistance, which is complete against avirulent races of the pathogen; and pseudomonogenic resistance is the common form of vertical resistance used against potato blight and the cereal rusts. There is, however, much incomplete (partial) vertical resistance, even against cereal rusts. It is commonly associated with intermediate reaction types, such as X, 2, or 3−. This incomplete resistance permits the fungus to sporulate, although on a reduced scale, and an epidemic can proceed, although at a reduced rate. Thus, slow rusting or slow mildewing of cereals can be vertical as well as horizontal. In Section 9.8 a test is suggested to distinguish between horizontal and vertical partial resistance. In the present section we are concerned only to show that if vertical resistance genes condition only partial resistance to avirulent races of the pathogen, vertical resistance can reduce the infection rate.

The data of Rees *et al.* (1979) recorded in Table 13.2 illustrated this. Experimental plots were established in Queensland of wheat cultivars with intermediate resistance to leaf rust caused by *Puccinia recondita tritici*. These were infected from artificially inoculated spreader rows, and the progress of disease recorded at successive dates. Two isolates of *P. recondita tritici* were used. To these isolates the wheat cultivars listed in Table 13.2 had an intermediate level of resistance. That is, the resistance was incomplete. The cultivars developed rust slowly; the infection rate was reduced. This can be seen from the area below the disease progress curve, relative to more resistant and more susceptible cultivars. In 1971 the area associated with the cultivars in Table 13.2 varied from 24 to 190 compared with 0 for a fully resistant cultivar (Transfer) and 2156 for the exceptionally susceptible cultivar Morocco. In 1972 the figures were 3 to 716, compared with 0 for Transfer and 1278 for Morocco. Ranked according to the area below the disease progress curve, the cultivars performed differently in the 2 years (i.e., against different isolates of the pathogen). Much of the resistance was therefore vertical.

TABLE 13.2

Slow Rusting by Two Races of *Puccinia recondita* of Wheat with Vertical Resistance[a]

Cultivar	Infection type		Adult reaction[b]		Area below curve[c]		Rank[d]	
	1971	1972	1971	1972	1971	1972	1971	1972
Warchief	12 +	—	MR–R		24	—	1	—
Hopps	0;2 +	X	MS–MR	MR–R	56	3	2	1
Gamenya	0;2 −	3 +	MR–R	S	63	370	3	6
Pusa 80-5b	2 + 3 + +	—	MS	—	86	—	4	—
Mengavi	0;2 −	3 +	MR–R	S	121	716	5	7
Dural	3 + +	3 − c	MR	MR–R	181	36	6	4
Kenya Governor	3 +	2	MS	MS–MR	186	121	7	5
Festival	2 + +	2 − n	S–MS	MR–R	190	4	8	2
Warput	—	3 −	—	MR–R	—	14	—	3

[a] Data of Rees *et al.* (1979). The race in 1971 was race 68-Anz-1,2,3,4; and in 1972 race 68-Anz-2,3. The dates 1971 and 1972 represent not only different years but also different races.

[b] R = resistant; MR = moderately resistant; MS = moderately susceptible; and S = susceptible.

[c] The area below the disease progress curve. At the final reading the incidence of leaf rust varied from 30 to 60% in 1972; data were not published for 1971.

[d] The ranking is according to the area below the disease progress curve.

Another example comes from the work of Clifford and Clothier (1974) on *P. hordei* in barley. The barley cultivar Vada has some resistance to rust, which develops more slowly than in susceptible cultivars. Clifford and Clothier showed that this resistance was at least partly vertical. Using as their criterion the production of spores per square centimeter of inoculated leaf surface, they demonstrated a highly significant differential interaction between barley cultivars and isolates of *P. hordei*, which identifies the resistance as vertical, with or without some horizontal resistance.

Much is being written these days about slow rusting (see Chapter 15) and slow mildewing of cereal crops, possibly on the mistaken assumption that because horizontal resistance reduces the infection rate (see Chapter 14), a low infection rate is inherently stable. If the slow infection rate comes from vertical resistance it has little to commend it; pseudomonogenic vertical resistance would almost always be better.

14

Epidemiological Effects of Horizontal Resistance

14.1 INTRODUCTION

This chapter deals with the effect of horizontal resistance on the progress of disease in fields. Of necessity, we discuss resistance that is only partial. The general effect of horizontal resistance is to reduce the infection rate within any set of environmental factors.

Resistance that is demonstrably horizontal is necessarily only partial. Complete resistance, which the pathogen cannot overcome even partially, cannot be identified as either horizontal or vertical. We may, e.g., suspect that the monogenic resistance of cabbage to *Fusarium oxysporum* f. sp. *conglutinans* is horizontal; but while this resistance remains complete there is no way of distinguishing whether it is horizontal or vertical, nor is there any point in trying to do so. A demonstration to distinguish main effects from interactions, which is what the classification of resistance as horizontal or vertical is about, needs at least two resistance host genotypes differing in their resistance and two pathogen genotypes differing in their pathogenicity. The need cannot be met in horizontal resistance unless it is partial; and complete horizontal resistance, which is probably common, is outside the scope of our discussion. All this does not mean that we equate partial resistance with horizontal resistance; partial resistance can also be vertical or a mixture of horizontal and vertical.

As in the previous chapter potato blight is used as the main vehicle of discussion. From about the mid 1930s potato breeders started to release cultivars with vertical R genes derived from *Solanum demissum*, but cultivars released before then were almost entirely horizontal in their resistance. Many of these old cultivars are still grown, and the resistance of some of them has been studied. Another reason for choosing potato blight for dis-cussion is that more quantitative information is available for potato blight than for most other diseases.

14.2   HISTORY OF BLIGHT RESISTANCE IN THREE POTATO VARIETIES

Table 14.1 lists three potato varieties: Bintje, Eigenheimer, and Voran. They have been assessed annually in the Netherlands for resistance to *Phytophthora infestans*, on a scale in which 3 = very susceptible and 10 = very resistant. Bintje with a rating of 3 can be considered to have practically no resistance; Eigenheimer is intermediate; and Voran with a rating of 7–8 is moderately resistant both in the foliage and the tubers. None of these varieties has a R gene, and their resistance, such as it is, has remained stable over a period when varieties with R genes have come and gone as blight resisters. The great stability of Voran's resistance despite great changes in virulence of *P. infestans* attacking varieties with R genes indicates the absence of noteworthy Voran × pathogen interaction and therefore of vertical resistance.

TABLE 14.1

Date of Introduction, and Assessments in 1938 and 1968 of Resistance
to *Phytophthora infestans* in the Foliage and Tubers
of Three Potato Cultivars in the Netherlands[a]

| | | Assessed resistance[c] | | | |
| | | Foliage | | Tubers | |
Cultivar[b]	Date of introduction	1938	1968	1938	1968
Eigenheimer	1893	4	5	4	3
Bintje	1910	3	3	5	3
Voran	1936	7	7	7	6.5

[a] Data from the Nederlandse Rassenlijst voor Landbouwgewassen.
[b] These cultivars have no R gene.
[c] 3 = very susceptible; 10 = very resistant.

Table 14.1 spans 30 years and shows how the three varieties were assessed in 1938 and 1968, these two dates marking the beginning and end of Voran's career as a cultivar in the Netherlands. Eigenheimer provides a useful check on any subjective changes in the assessments. By 1938 Eigenheimer had already been a popular variety for 45 years. It had been widely grown because of its good yield and outstanding flavor; and it had been widely grown in that part of the Netherlands where Voran later established itself. During those 45 years host and pathogen had presumably reached an equilibrium, and subsequent changes probably mark the limit of subjective change among assessors. Between 1938 and 1968 Eigenheimer's assessment changed very little; the resistance of the foliage was rated slightly higher and that of the tubers slightly lower in 1968 than in 1938. Bintje, too, was an old variety in 1938, and its evidence supports Eigenheimer's. With both these varieties there is some evidence that by 1968 assessors had raised their standards for tuber resistance.

Voran's assessed resitance remained substantially stable during its career of about 30 years, despite great changes in the area of its cultivation. Table 14.2 gives some details about Voran. It was registered as a cultivar in the Netherlands in 1936. In 1937 it was still a young cultivar covering only a small area. Cultivation was mainly near fields of the old established cultivar Eigenheimer, and it was presumably from blighted Eigenheimer that fields of Voran received inoculum of *P. infestans*. Thereafter, Voran's popularity increased, and it became a favorite on the reclaimed peat soils of the northern

TABLE 14.2

Resistance to *Phytophthora infestans* in the Potato Cultivar Voran Assessed at Intervals from 1937 through 1970[a]

Date	Assessed resistance		Status of cultivar
	Foliage	Tubers	
1937	6	7	Rare; new introduction
1938	7	7	
1942	8	7	
1946	8	7	
1950	8	7	Dominant over a large compact area
1952	8	7	
1957	7	6.5	
1960	7	6.5	
1968	7	6.5	
1970	7	6.5	Cultivar practically extinct

[a] See the footnotes in Table 14.1.

Netherlands. In 1950 it reached a peak of popularity, covering nearly 80% of the 40,000 hectares of potatoes in an area of 80 × 20 km. Later, it was replaced by newer cultivars and by 1970 was almost extinct. Voran's history was thus one of great change, from near insignificance in 1937 to dominance in 1950, and then back again to insignificance in 1970. In 1937 inoculum of *P. infestans* must have come from blighted fields of cultivars other than Voran. In 1950 it was largely self-contained for inoculum, Voran fields receiving inoculum from blighted Voran. In 1970 blighted Voran fields were an insignificant source. Yet the assessment of Voran's resistance remained unchanged. Had there been a host × pathogen interaction with *P. infestans* becoming adapted to Voran, Voran would have lost resistance between 1937 and 1950 and regained it between 1950 and 1970. There was no such trend. Table 14.2 shows that what resistance Voran had in the beginning, it kept; and the evidence that Table 14.2 summarizes is far more massive and relevant than any evidence that could have been extracted from puny artifacts in the greenhouse or laboratory.

According to van der Zaag (1956) the foliage of Voran has some resistance to infection. He dipped leaflets in a dilute suspension of zoospores of *P. infestans*. Fewer lesions were formed in Voran than in varieties regarded as susceptible.

14.3   HORIZONTAL RESISTANCE AND A REDUCED INFECTION RATE

Partial resistance, regardless of whether it is horizontal or vertical, reduces the infection rate. In this chapter we are concerned only with horizontal resistance.

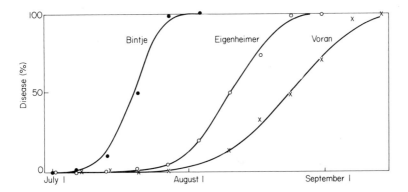

Fig. 14.1. The progress of blight in 117 fields of three potato varieties. The data are for the sand area of the Netherlands in 1953 (Anonymous, 1954). From Vanderplank (1968, p. 30).

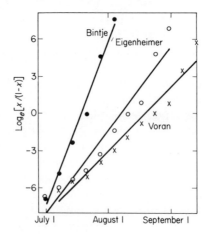

Fig. 14.2.   The progress of blight in 117 fields of potato varieties. The data are the same as in Fig. 14.1, with the proportion x of foliage diseased converted into $\log_e[x/(1 - x)]$.

Figure 14.1 compares the three potato varieties Bintje, Eigenheimer, and Voran in another way, by means of blight progress curves. The data are for 117 fields of potatoes in the sand area of the Netherlands in 1953 (Anonymous, 1954). The fields were not treated with a fungicide. The Netherlands system of blight assessment was interpreted by Cox and Large's (1960) conversion curve.

For each variety the blight progress curve is roughly S-shaped, and the curves reflect the differences in susceptibility. Bintje was completely blighted by the end of July, Eigenheimer by the end of August, whereas Voran carried on into September.

Differences are also reflected in the infection rates, and the appropriately converted data are shown in Fig. 14.2. The regression coefficient of $\log_e[x/(1 - x)]$ on time measures the infection rate r. From Fig. 14.2 we estimate r to be 0.42, 0.21, and 0.16 per day for Bintje, Eigenheimer, and Voran, respectively. These are gross averages over the whole period during which blight was observed. They do not reflect known ontogenic effects or possible weather changes.

14.4   COMPONENTS OF HORIZONTAL RESISTANCE

Horizontal resistance to potato blight manifests itself in different ways:

1. Plants resist infection. When plants are inoculated with the same number of spores, fewer lesions are formed on plants of a resistant variety.

2. Sporulation is less abundant in lesions on a resistant variety than on a susceptible one.

3. From the time of inoculation it takes longer for sporulation to start on a resistant variety than on a susceptible one. The latent period is longer.

4. Infected tissue ceases to be infectious sooner. The infectious period is shorter. In blight lesions on potato leaves an infertile zone follows behind the sporulating zone and reduces the width of lesions actively producing spores.

Resistance to infection was recorded by Müller (1953). He sprayed a dilute suspension of zoospores of *P. infestans* on the upper surface of discs cut from leaves of different potato varieties. The proportion of discs that became infected was less when the discs came from varieties regarded as field resistant. Hodgson (1961) used the same technique with the same results. Van der Zaag (1959) inoculated leaflets by dipping them in a dilute zoospore suspension. Fewer lesions were formed in varieties regarded as field resistant. Lapwood (1961) found differences between varieties in the resistance of the leaf lamina to infection, but thought other differences more important. Bigger differences were found when he inoculated varieties in the leaf axil and counted the number of axillary buds infected and destroyed.

Phytophthora infestans advances more slowly through the tissues of resistant potato varieties. Lapwood (1961) found important differences in the rate and extent of the advance through petioles and stems. In the resistant

TABLE 14.3

**Width of the Sporing Zone of Potato Blight Lesions in Two
Varieties at Three Different Heights in the Canopy[a]**

Year and position in the canopy	Width of sporing zone (mm) in:	
	Up-to-Date	Arran Viking
1957		
Upper	2.7	1.6
Middle	3.9	2.6
Lower	4.4	3.0
1958		
Upper	4.6	3.5
Middle	5.0	4.2
Lower	5.5	4.5

[a] Data of Lapwood (1961), from Vanderplank (1963, p. 256). Figures are the means of 26 days in 1957 and 12 days in 1958. The weather in 1958 was unusually wet.

variety Arran Viking the lesions usually remained small. In the susceptible varieties Up-to-Date and King Edward the lesions often girdled the leaf petiole, the leaf collapsed, and after a while the stem was also girdled.

Phytophthora infestans sporulates more abundantly on susceptible potato varieties, and this is related to less necrosis in the lesions in these varieties (Kammermann, 1950; Lapwood, 1961). Lapwood made a detailed study of this. The sporing zone in a lesion surrounds a necrotic area. Lapwood marked the progress of the sporing zone. In susceptible varieties in wet weather the zone was continuous from day to day. That is, the zone was 1 day's growth wide. In resistant varieties faster necrosis reduced the sporing zone. In Table 14.3 the sporing zones of the susceptible variety Up-to-Date and the more resistant variety Arran Viking are compared. In 1958, a "blight year," the zone was one-fifth less for Arran Viking than for Up-to-Date, although the rate of advance of the lesion was about the same.

The latent period, the time the pathogen takes to start sporing after inoculation, is longer in resistant varieties. Vowinckel (1926), Kammermann (1950), Schaper (1951), Rudorf and Schaper (1951), and Lapwood (1961) studied *P. infestans* in potato varieties. Under favorable conditions sporulation is abundant after 4 days in susceptible varieties, but 5 or 6 days are needed in resistant varieties. Table 14.4 records some data of Lapwood (1961); they show how sporulation is delayed in resistant varieties.

TABLE 14.4

Varietal Difference in the Latent Period of Potato Blight[a]

	Days after inoculation				
Variety	3	4	5	6	7
Up-to-Date	0	12	23	24	—
King Edward	0	8	22	24	—
Majestic	0	14	21	24	—
Ackersegen	0	0	18	23	24
Ås	0	1	13	20	23
Ontario	0	3	18	22	24
Average susceptible	0	11	22	24	—
Average resistant	0	1	16	22	24

[a] Data of Lapwood (1961) for droplet inoculations, from Vanderplank (1963, p. 254). Figures are the number of discs out of 24 that were sporing.

14.5   ONTOGENIC EFFECTS

The age of a plant affects its resistance to disease. Ontogeny affects all components of resistance except, possibly, resistance to infection, i.e., the proportion of spores that establish lesions.

Horizontal resistance to potato blight is greatest in middle age, and plants become more susceptible as they grow older. This was observed by de Bary (1876), Müller (1931), Grainger (1956), and others. Middle-age resistance is clearest in late-maturing varieties. This can be seen in Fig. 14.2. Infection in the three varieties was observed early in July. Bintje, which is early maturing, succumbed without delay. In Eigenheimer and Voran, in the same weather conditions, the infection rate, measured as the regression of $\log_e[x/(1 - x)]$ on time, stayed low during July and then increased as the varieties approached maturity.

Other diseases have other patterns. In powdery mildew of barley caused by *Erysiphe graminis hordei* young-plant susceptibility followed by middle-age resistance is conspicuous in northwestern Europe; and the disease progress curve begins steeply.

The potato blight pattern, with middle-age horizontal resistance followed by old-age susceptibility, is common. Grainger (1968) records examples. It is important because it possibly links up with sink-induced susceptibility (see Chapter 11) and the vertifolia effect. There is a large, almost unexplored field of study here; and the study of disease progress curves may help define the extent of the conflict between breeding for horizontal resistance and breeding for the highest potential yield in the absence of disease.

15

Slow Rusting of
Cereal Crops

15.1 INTRODUCTION

Slow rusting of wheat, barley, and oats has been much discussed in the recent literature. The name, slow rusting, seems to have been first used by Caldwell *et al.* (1970) for resistance to wheat leaf rust, but the phenomenon under various names such as tolerance, generalized resistance, general resistance, field resistance, and adult- (mature-) plant resistance has been studied for many years. Slow rusting is partial resistance partitioned in variable proportions between horizontal and vertical resistance. Vertical slow-rusting resistance has already been discussed in Section 13.14. It needs no more than passing reference in this chapter. The emphasis in slow-rusting research has been on the horizontal component because the purpose of using slow-rusting cultivars is to substitute the stability of horizontal resistance for the instability that has often been such an undesirable feature of vertical resistance.

There is nothing new about slow-rusting resistance other than its name. In agricultural practice, it is the resistance that has been used against maize rust all along. Slow-rusting cultivars of durum wheat have been adequately resistant to leaf rust in North Dakota (Statler *et al.*, 1977a), and the literature abounds with references to cultivars of wheat, oats, and barley that rust slowly. Nor is it new in agricultural theory. The findings of Gassner and

Kirchhoff (1934) with *Puccinia graminis tritici* and *P. recondita tritici* in wheat, *P. coronata* in oats, and *P. hordei* in barley, and of Vohl (1938) with *P. recondita tritici* reveal concepts more in advance of those in much of the recent literature.

There is nothing special about slow-rusting resistance. In its essential features it resembles slow blighting of potatoes, and *mutatis mutandis* Chapter 15 is little more than an extension of Chapter 14. Ominously, there are hints that there is a sink-induced loss of slow-rusting resistance, and that Chapter 15 is an extension of Chapter 11.

Caldwell *et al.* (1970), Verna *et al.* (1975), Ohm and Shaner (1976), Statler *et al.* (1977a,b), Gavinlertvatana and Wilcoxson (1978), Kuhn *et al.* (1978), Shaner and Hess (1978), Shaner *et al.* (1978), and Tomerlin *et al.* (1983) have discussed slow leaf-rusting of wheat by *P. recondita*; Martin *et al.* (1977, 1979), Rowell and McVey (1974), and Skovmand *et al.* (1978) have discussed slow stem-rusting of wheat by *P. graminis tritici*; Russell (1976) has discussed slow stripe-rusting of wheat by *P. striiformis*; Simons and Michel (1968), Simons (1969, 1975), Heagle and Moore (1970), Luke *et al.* (1972, 1975a,b), and Kochman and Brown (1975) have discussed slow crown-rusting of oats by *P. coronata*; Krull *et al.* (1965), Kochman and Brown (1975), and Sztejnberg and Wahl (1976) have discussed slow stem-rusting of oats by *P. graminis avenae*; and Clifford (1972, 1975), Parlevliet (1976), and Johnson and Wilcoxson (1978) have discussed slow rusting of barley by *P. hordei*. In this literature there are many points of agreement.

1. The thesis that stable horizontal resistance to the cereal rusts exists is valid. The evidence is of the same sort as that for the blight resistance of the potato cultivar Voran discussed in the previous chapter: There is an historic record of stability. Wheat, barley, and oat varieties are known with slow-rusting resistance that has remained stable over the years, apparently free from the boom-and-bust cycles that have plagued varieties with vertical resistance.

2. Breeding for slow-rusting resistance is feasible. The heritability of resistance is great, heritability values of 80% or more having been recorded.

3. Resistance is continuously distributed between parent varieties and between progeny in the F_2 and later generations. Transgressive inheritance occurs, and resistance can be accumulated beyond that of the parents.

4. Resistance is conditioned by relatively few genes, estimates of gene pair numbers varying from less than 2 to more than 20, but usually within the range of 2 to 6.

5. Resistance is inherited additively. Dominance and epistasis effects have been reported, but they are small and perhaps unreal in light of the difficulty of estimating true midpoints of disease (see Section 3.9).

6. The components of the horizontal resistance are the same as those discussed for potato blight in Chapter 14, namely, a reduced proportion of spores that manage to establish lesions, smaller lesions or lesions with reduced spore production, and a longer latent period. Except for a longer latent period, these add up to a lower progeny/parent ratio, to be discussed in the next chapter.

7. The area under the disease progress curve has been widely used to measure whether rusting was fast or slow.

15.2 LATE RUSTING AND SLOW RUSTING

A smaller area under the disease progress curve may result either from late rusting or from slow rusting. These are terms distinguished by Luke *et al.* (1972) in relation to crown rust of oats. Some late-maturing oat cultivars, particularly Red Rustproof (*Avena byzantina*), exhibited both late rusting and slow rusting. Late rusting refers to the start of the rusting, and commonly reflects vertical resistance (see Chapter 13). But late rusting also reflects middle-age horizontal resistance, analogous to the late blighting of the potato cultivar Voran, which, it will be remembered from Chapter 14, blighted in August and September, whereas the susceptible cultivar Bintje blighted in July. Slow rusting means that the infection rate, once rusting has begun, is low. As was pointed out in the previous section, slow rusting can be vertical or horizontal, but the emphasis is on horizontal resistance.

15.3 MAIZE RUST

Slow rusting of maize by *P. sorghi* is the prototype of slow rusting in cereal crops, although the name slow rusting does not seem to have been used.

Slow rusting of maize is mostly horizontal, although some maize lines have much vertical resistance given by *Rp* genes. Hooker and Le Roux (1957) tested a range of maize lines against isolates of *P. sorghi* obtained as uredial collections from widely separated locations in North and Central America or as aecial collections from infected *Oxalis* spp.. Their results illustrate how variable is the mixture of horizontal and vertical resistance. The top five entries in Table 15.1 are for maize lines with great vertical resistance, the bottom five for lines with little vertical resistance. If one confines attention to the North American maize belt, resistance is mostly horizontal, vertical resistance seemingly being almost negligible. Hooker and Le Roux (1957) tested 85 maize lines from Iowa and Wisconsin. Only

TABLE 15.1

Proportion of Isolates of *Puccinia sorghi* to Which Some
Maize Lines Were Resistant[a]

Maize line	Origin	Proportion of avirulent isolates
Cuzco	Peru	59/59
Leon 1 27-4-1	Mexico	30/38
GG 208 R	USA	43/44
B 38	USA	39/44
K 148	USA	37/44
Midland 24-2-1	USA	1/16
(Oh45 × W92)-1-1-2	USA	1/17
Pop 35	USA	1/17
L 317	USA	0
B 14	USA	0

[a] From data of Hooker and Le Roux (1957). Thus, Cuzco was resistant to all the isolates of *P. sorghi*, L 317 and B 14 to none of them. Cuzco had high vertical resistance and presumably many *Rp* genes. L 317 and B 14 had little or no vertical resistance and presumably few or no *Rp* genes.

1 of these lines had seedling resistance, and that resistance was to only 2 of the 15 cultures of *P. sorghi* that Hooker and Le Roux used. Yet of these 85 lines 16 were slow rusters in the field.

The inheritance of slow rusting in maize has been studied (Hooker, 1962, 1967a,b, 1979; Kim and Brewbaker, 1977). Parents and segregating F_2 populations of susceptible × susceptible, resistant × susceptible, and resistant × resistant crosses were scored for rust reaction. There was wide variation in reaction among plants in the F_2 generation. The variation ranged mostly from the rust scores of the one parent to those of the other, but sometimes there was transgressive resistance. Variation among plants of the F_2 generation of each cross was continuous, and the frequency distribution of individual plant scores was approximately normal. The mean score of each F_2 population approached the calculated average of the two inbred parents. The score of the F_1 generation was usually nearer that of the resistant than the susceptible parent. This may indicate partial dominance for resistance, or heterosis, or simply an error in the estimated mean score (see Section 3.9). Heritability estimates averaged 80–87%, according to the method of estimation. By selecting the most resistant 5% of the plants and using resistant inbreds as parents, genetic advance was swift, averaging 1.73 units per generation on a scale 1 = resistant and 7 = very susceptible. Resistance seems to be conditioned by relatively few genes, the resistant inbred

Oh545 being estimated to have two gene pairs conditioning resistance, and another resistant inbred Cm105 to have 1.3 gene pairs, this figure being the average obtained by three different methods of estimation. (See also Section 3.6.) It should not be difficult to select for slow rusting in a breeding program.

Experience has shown the resistance of slow-rusting maize inbreds and hybrids to be stable. The resistance accumulated in the early years is still effective.

15.4   RESISTANCE BEFORE AND SUSCEPTIBILITY AFTER FLOWERING

There is a marked ontogenic effect in slow-rusting resistance. As with horizontal resistance to potato blight (see Chapter 14), horizontal resistance to cereal rusts increases to a maximum and then decreases. High horizontal resistance developed up to the time of flowering (anthesis) is lost, at least partially, after flowering.

Gassner and Kirchhoff (1934) worked with what we now call slow rusting. They noted that against *P. graminis* and *P. recondita* in wheat, *P. coronata* in oats, and *P. hordei* in barley there was a wave of resistance just before tillering and flowering, followed by a wave of susceptibility in old plants and older parts of plants. The wave of susceptibility in older plants was first seen in the middle of the stems and later in the top leaves. Gassner and Kirchhoff coined the name old-age susceptibility. Vohl (1938), working with *P. recondita*, confirmed Gassner and Kirchhoff's findings. There is a typical rhythm of field resistance in the common wheat cultivars Marquis, Marquillo, Thatcher, Hope, H44, Garnet, and Reward. Plants of these varieties are susceptible in the seedling stage, begin to become resistant in the shooting stage, are resistant at the time of heading and flowering, and are finally susceptible again in old age.

Caldwell *et al.* (1970) reported that the winter wheat cultivar Vigo, grown on nearly two million acres in 1954, has remained free from severe leaf rust in pure stands in Indiana, despite its being highly susceptible at maturity. It may become severely infected after senescence starts. They also reported that some spring wheat cultivars have shown promising levels of slow rusting at presenescent stages in the plant's growth, even under heavy inoculum. A pattern of infection, wherein early pustules occur mainly on the basal 10–20% of the blade, was found in the cultivar Mentana and its derivatives Lerma 50 and 52. With senescence in the host, pustules spread to the distal portion. Such resistance to infection in the cultivar Lerma Rojo 64 and its derivatives gave significant protection to commercial fields in Mexico.

Ohm and Shaner (1976), also working with wheat leaf rust, were more precise about the plant's growth stage (GS). Resistance, judged by reduced pustule size, was greatest when plants were inoculated between GS 39 (flag-leaf emergence) and GS 60 (flowering) and least when inoculated after GS 68. Ohm and Shaner point out that even though slow-rusting resistance declines in old age, it may nevertheless provide enough protection to the crop before and during flowering to prevent significant loss of yield from leaf rust. Shaner and Hess (1978), working on equations for integrating components of slow-rusting resistance to leaf rust in wheat, comment on the large increase in severity of leaf rust often seen in slow-rusting wheats just before ripening.

When slow rusting is discussed in the literature as mature- (adult-) plant resistance, maturity is usually not defined. Mature plants are simply taken to exclude seedlings. When maturity is defined, it seems to end with the development of the flag leaf or with flowering. With this agreed, the evidence is consistent. There is resistance before flowering and loss of resistance after flowering. Some may prefer the names, preanthesis resistance and post-anthesis susceptibility.

15.5   SINK-ASSOCIATED AND SINK-INDUCED LOSS OF RESISTANCE

Why stress flowering as a turning point? The grain sink starts with flowering. After-flowering loss of resistance is sink-associated loss of resistance. The evidence for after-flowering loss of resistance in the previous section is *ipso facto* evidence for sink-associated loss of resistance.

Sink-associated loss of resistance gives a hint that there may be sink-induced loss of resistance. Is there? Relevant literature is scarce. Simons (1975) worked with crown rust of oats caused by *P. coronata*. The very susceptible oat cultivar Clinton was crossed with 4 lines of slow-rusting oats. Segregating lines in the F_2, F_3, and F_4 generations were selected for adaptation, with emphasis on maturity. In the F_5, F_6, and, sometimes, the F_7 generations 100 adapted lines from each cross were tested for correlation between yield potential in the absence of rust and resistance to rust. All correlations except one were statistically significant, usually highly significant ($P < .01$), and in all the significant correlations high resistance was associated with lower yield potential. For example, consider the cross between Clinton and the slow ruster P.I. 174544 which was agronomically adapted. In the F_5 generation the correlation coefficient between resistance and yield potential was -0.421 and in the F_6 or F_7 generation -0.692. Of the 10 highest yielding lines from the Clinton × P.I. 174544 cross 3 were significantly more susceptible than Clinton, 4 less susceptible than Clinton, and all less resistant than P.I. 174544. None of the lines from any of the crosses combined the highest yield potential with maximum resistance.

This is evidence for a sink-induced loss of resistance, but could also be interpreted as evidence that resistance is linked to poor yield. The linkage theory does not satisfactorily explain the reversal at flowering, from resistance before flowering to susceptibility afterwards, but cannot be ignored. More direct evidence with the linkage factor excluded is needed, as when it is determined what happens to resistance if the flowers are cut off.

One cannot exclude the possibility of a sink-induced loss of horizontal, slow-rusting resistance, with lower yields the price to pay for higher resistance. On the evidence of maize, the price (if there is a price) is not necessarily too high; but it would be imprudent to generalize blindly and extrapolate from maize to other grain crops. More information is needed.

15.6   BREEDING FOR HORIZONTAL SLOW-RUSTING RESISTANCE

Heritability of slow-rusting resistance is high; a major part of the resistance is controlled by few genes; and the evidence is for a quick gain of resistance in progenies under appropriate selection pressure. The key difficulty is to arrange for appropriate selection pressure. Selection for horizontal resistance must be isolated from complicating vertical resistance. When parents contain, as they commonly do, R genes (Sr, Lr, Yr, Pg, Rp genes, and their like) the vertical resistance that they give must be excluded during selection.

Vanderplank (1963), thinking mainly of potato blight, pointed out that uniform selection for horizontal resistance is possible only if progenies are exposed to a single race of the pathogen, that race being virulent for all the R genes. Thus, if the breeding material contained genes $R1$, $R2$, $R3$, and $R4$, only race (1, 2, 3, 4) or another race with the numbers 1, 2, 3, and 4 in its designation would subject progenies to equal selection pressure for horizontal resistance. A mixture of races, e.g., a mixture of races (1, 2, 3), (1, 2, 4), (1, 3, 4), and (2, 3, 4), as well as lesser races, would mean that there would be unequal initial inoculum for different host genotypes. The condition that the inoculum should be of a race virulent for all the R genes is apt enough for screening potatoes against blight, but inept in the selection of cereal crops against rust because wheat, barley, and oat plants usually have mainly unidentified R genes.

Robinson (1976) also stressed the need for using only a single race of the pathogen in screening parents and progenies. Potential parents are challenged by a single race, those that show strong resistance that might be vertical are discarded, crosses are made, and progenies throughout the successive generations are challenged by the same single race. Robinson's method is widely applicable, and any R genes that might be present need not be known. The reasoning is clear. Suppose that race (1) of the pathogen was used as inoculum. The only accepted parents would be those with the gene $R1$ or with no R gene at all. Potential parents with the genes $R2$, $R3$, etc. would be discarded at the start. Progenies in the F_1 and subsequent generations might well have the gene $R1$, but because race (1) was used throughout, there could be no vertical resistance. So too if race (2) was used as inoculum, the parents, the F_1 generation, and subsequent generations might all have the gene $R2$, but there would be no vertical resistance. But suppose that a mixture of races (1) and (2) was used. Parents with the gene $R1$ alone or $R2$ alone would show no vertical resistance and would be accepted. But progenies that combined the genes $R1$ and $R2$ would be vertically resistant. The use of a mixture of races opens the way to transgressive vertical resistance and would vitiate the selection of horizontal resistance.

Robinson's test is not perfect because it would allow vertical resistance to accumulate through successive generations if the R genes in the parents conditioned only weak vertical resistance associated, e.g., with a 3 or 3 − reaction type. The buildup of resistance would not necessarily always be horizontal even if the precaution was taken of using a single pathogenic race throughout. See Section 7.7 for cumulative vertical resistance. The test suggested in Section 9.8 might solve the problem of ascertaining whether the built-up resistance was horizontal or vertical.

16

Resistance against Endemic Disease

16.1 INTRODUCTION

Epidemic and endemic disease differ in the way resistance most effectively controls them. A change from epidemic to endemic disease reduces the value of a long latent period relative to other components of resistance and reduces the value of vertical relative to horizontal resistance. To control endemic disease, emphasize the role of a reduced progeny/parent ratio and of horizontal resistance.

To illustrate epidemic and endemic disease, compare potato blight caused by *Phytophthora infestans* with leaf blight of taro caused by *P. colocasiae*. Potato blight is typically an epidemic disease. Seasonal epidemics begin from small beginnings. One infected shoot from an infected seed tuber per square kilometer of potato fields is more than enough to start an epidemic (van der Zaag, 1956). Most fields of potatoes begin healthily and only become engulfed by the blight epidemic later. Epidemics commonly end with the destruction of all foliage, which means several billions of lesions per square kilometer; during the course of the epidemic *P. infestans* can increase a billionfold or more. Epidemics are erratic in their occurrence; they occur in "blight years," when the weather favors them.

Leaf blight of taro (*Colocasia esculenta*) caused by *P. colocasiae* is an endemic disease of the humid tropics. Putter (1980) has described it as an endemic disease in Papua New Guinea. The taro is an annual root crop and in Papua New Guinea is cultivated throughout the year, being harvested and replanted continuously. In any one garden taro leaves are always present. So too is the fungus. There are variations in the level of disease, but no period is free from disease. From January to May, higher rainfall and cloudiness raise the level of disease. From June to December, lower rainfall lowers the level. In contrast with potato blight, inoculum is always present; all gardens have the disease; seasonal increases in the fungus population are restricted; and seasonal variation in disease incidence is small.

The differences between epidemic and endemic disease, illustrated above, have long been known in medicine and have been embodied in definitions.*

Epidemic disease, according to Webster's New International Dictionary, tends to affect many persons within a community, area, or region at one time. The example the dictionary gives is that many children died that winter of epidemic fevers. The Oxford English Dictionary says that epidemic disease is prevalent among a people or community at a special time and is produced by some special causes not generally present in the affected locality.

Endemic disease, defined by Webster's Dictionary, is disease peculiar to a locality or region, constantly present to a greater or lesser extent in a particular place, and distinguished from epidemic or sporadic disease. The Oxford Dictionary defines endemic disease as disease habitually prevalent in a certain country and due to local permanent causes.

Mutatis mutandis, these definitions established in medicine apply well in plant pathology. Epidemic disease is sporadic, occurring within limited time, occasionally, and produced by special causes such as special weather (hence disease forecasting) or the development of virulence in a hitherto avirulent pathogen population. Endemic disease is constantly present, as when one can predict simply from past experience that a particular disease is likely to occur during the coming season.

The classification of disease as either epidemic or endemic puts all disease into two categories. There must inevitably be much overlapping and

---

* Some plant pathologists toy with the terms, r-strategy and K-strategy, when they write of epidemic and endemic disease. This has little to commend it. There is a convention involved of giving precedence to previous usage. When MacArthur and Wilson (1967) introduced the terms "r-selected" and "K-selected" species, they used concepts about reproductive strategies already familiar to epidemiologists both in medicine and plant pathology. Ecologists' belated awakening to what epidemiologists already knew is no reason why epidemiology should now be dressed in ecological clothing. Admittedly, plant pathology has an ecological side, but its affinity with medicine is far greater. Think of the everyday words in plant pathology derived from medicine: disease, infection, inoculation, antigen, ... It is not disputed that epidemiology can gainfully adopt ideas from ecology, as ecology can gainfully adopt ideas from epidemiology. But this note is about terminology, not ideas.

smudging of distinctions. Two categories cannot exactly cover every possibility in the whole range of progeny/parent ratios from 1 to ∞, with the ratios and latent periods varying from day to day and hour to hour. Short of having an infinite number of categories one must put up with occasional difficulties of classification. But if for obvious reasons it is desirable to have the fewest possible categories, then the two, epidemic and endemic disease, seem to be the most useful. They reflect medical experience; they are relevant in plant pathology; and they can be linked to progeny/parent ratios.

16.2   THE PROGENY/PARENT RATIO

Mathematically, the difference between epidemic and endemic disease is simple and direct. In epidemic disease, the progeny/parent ratio is great; in endemic disease, it is small, in the neighborhood of 1 and probably not exceeding 5. By the progeny/parent ratio is meant the average number of daughter lesions established directly by the spores released from a single parent lesion during the parent lesion's whole lifetime. In terms of systemic disease, it is the average number of systemically diseased plants infected by inoculum from a single, diseased parent plant. The ratio is meant to apply to low levels of disease when the amount of healthy tissue (or in systemic disease the number of healthy plants) can be regarded as unlimited, i.e., when no allowance need be made for multiple infection.

The effect of the progeny/parent ratio is illustrated by Table 16.1, which is calculated from a simple model discussed in an appendix to this chapter. The

TABLE 16.1

Upper Limit of Disease in Relation to the Initial Inoculum and the Progeny/Parent Ratio[a]

Initial inoculum[b]	Progeny/parent ratio[c]								
	0	0.5	1.2	1.5	2.0	3.0	4.0	5.0	10.0
<0.00001			0.314	0.582	0.797	0.940	0.980	0.993	>0.999
0.001	0.001	0.002	0.318	0.584	0.797	0.941	0.980	0.993	>0.999
0.01	0.01	0.020	0.348	0.594	0.800	0.941	0.980	0.993	>0.999
0.1	0.1	0.176	0.514	0.678	0.828	0.948	0.982	0.994	>0.999
0.3	0.3	0.437	0.697	0.784	0.879	0.961	0.986	0.995	>0.999

[a] The upper limit $L = 1 - (1 - C) \exp(-\alpha L)$. It is the proportion of disease reached, given unlimited time, when disease proceeds according to Eq. (16.1).

[b] This is $C = x(0)$, the constant of integration in Eq. (16.1). The initial inoculum is expressed as the initial proportion of disease it causes or represents.

[c] This is α in Eq. (16.1).

table shows how the progeny/parent ratio determines the upper limit of disease which starts from different levels of initial inoculum. The entries in the top row of the table are for disease developing from very low levels of initial inoculum (relevant, e.g., to potato blight epidemics starting from one infected shoot or less per square kilometer of potato fields). The figure given in the table for initial inoculum is less than 0.00001, i.e., less than 0.001% disease. If the progeny/parent ratio is 5 or more, this very small amount of initial inoculum is enough to start an epidemic which, given time, will destroy more than 99% of the foliage. For all practical purposes, the epidemic, given time, can run to completion. With progeny/parent ratios less than 5 there is an upper limit to disease, even when initial inoculum is abundant. Consider, e.g., a progeny/parent ratio of 2. From very low levels of initial inoculum (in the top row of entries) the upper limit is 0.797, i.e., 79.7% disease. Increasing the amount of initial inoculum to 0.3, i.e., 30% disease, raises the upper limit only as far as 0.879, i.e., 87.9% disease. A very great increase of initial inoculum (a more than 30,000-fold increase of initial inoculum from very small to 30%) raises the upper limit of disease only from 79.7 to 87.9%. A restricted upper limit of disease despite much initial inoculum is a feature of all progeny/parent ratios below about 5. The progeny/parent ratio, more than the initial inoculum, determines the upper level of disease. If one defines endemic disease as having a progeny/parent ratio less than 5, it follows that endemic disease varies only between restricted limits. The lower limit, which is the initial inoculum, is not very small, by virtue of the nature of endemic disease; and the upper limit is held down by the low progeny/parent ratio.

Table 16.1 assumes unlimited time for change. In reality time is limited. For example, if one considers change within four generations (and ignores multiple infection) disease would increase 16-fold if the progeny/parent ratio were 2, but 160,000-fold if the ratio were 20. For obvious reasons, endemic disease can vary only within restricted limits. Variation in endemic disease is even more restricted than Table 16.1 suggests.

16.3 INFECTION RATES AND DISEASE LEVELS

Epidemic disease deals with infection rates, endemic disease with disease levels. Time is a dominant element in epidemic disease, hence infection rates. Time is a minor and erratic element in endemic disease, hence disease levels that time only erratically affects. Time is measured by the latent period which determines the length of a generation. The latent period is therefore a major factor in epidemic disease, a minor factor in endemic disease. Observe that the latent period does not enter Table 16.1.

The average level of endemic disease is set by the progeny/parent ratio. In long-established endemic disease the initial inoculum recedes to become an insubstantial memory, and the progeny/parent ratio remains as the surviving factor that decides disease levels.

The progeny/parent ratio enters epidemic as well as endemic disease. It enters via the infection rate, which is determined both by the latent period and the progeny/parent ratio. Whereas the latent period measures the interval between generations, the progeny/parent ratio measures the multiplication of disease within a generation. High ratios of 50 or more have been recorded in epidemic disease.

16.4 THE PROGENY/PARENT RATIO AND LATENT PERIOD IN THE STRATEGY OF USING DISEASE RESISTANCE

In breeding for resistance against epidemic disease, aim both at reducing the progeny/parent ratio and at lengthening the latent period. In breeding for resistance against endemic disease, concentrate on reducing the progeny/parent ratio and consider a lengthening of the latent period as relatively unimportant.

Against epidemic disease a small progeny/parent ratio and a long latent period are both important, and their relative importance varies with the

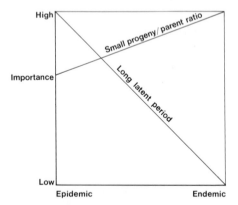

Fig. 16.1. The relative importance of a small progeny/parent ratio and a long latent period as components of resistance against epidemic and endemic disease. Epidemic disease and endemic disease are shown as extremes in a continuum. Thus, a long latent period is important in resistance against epidemic disease, unimportant against endemic disease, and of intermediate importance against disease between these extremes. The progeny/parent ratio reflects the proportion of spores that form lesions, the rate of spore production in lesions, and the period the lesion remains infectious.

infection rate. The faster the rate the more important, relatively, a long latent period becomes. See Vanderplank (1963, pp. 253–256).

Against endemic disease the latent period becomes important only in fluctuations, and here a long latent period is a two-edged weapon. A long latent period helps to retard upward fluctuations of disease, as during a spell of weather favorable to disease, but almost equally it hinders the subsequent downward fluctuation when the favorable spell is over.

Figure 16.1 illustrates how the progeny/parent ratio and the latent period differ in importance in epidemic and endemic disease. In this figure, epidemic and endemic disease are taken as extremes, with an indefinite number of intermediate states between.

Although the progeny/parent ratio and the latent period are epidemiologically distinct, they may be biochemically and physiologically related. Commonly, the one goes with the other (see Chapter 15). The relation may be inherent or simply the result of parallel selection.

16.5   HORIZONTAL AND VERTICAL RESISTANCE IN THE STRATEGY OF USING DISEASE RESISTANCE

In breeding for resistance to epidemic disease, both horizontal and vertical resistance have a place. In breeding for resistance against endemic disease, vertical resistance is of doubtful value, horizontal resistance of great value.

Endemic disease is insensitive to changes in the initial inoculum (see Section 16.2), but very sensitive to changes in the progeny/parent ratio. As a sufficiently accurate generalization, vertical resistance reduces the initial inoculum (see Chapter 13), and horizontal resistance the progeny/parent ratio (plus its effect on the latent period).

16.6   ENDEMIC DISEASE IN THE TROPICS

It is an oversimplification to say that plant disease in temperate zones is commonly epidemic and in tropical zones endemic. There is much endemic disease in temperate climates, and the tropics have witnessed major epidemics like this and last century's great coffee rust epidemics caused by *Hemileia vastatrix*. Also, tropical diseases that are as a rule essentially endemic can flare up into what the affected farmers must regard as epidemics. Nevertheless, endemicity is much more of a feature of plant disease in the tropics than in temperate climates. The reason is obvious. The seasons are less sharply divided in tropical climates than in temperate, and there is more continuity of disease.

Of special importance is the intertropical convergence zone, which is an area of generally low pressure and high rainfall that moves to and fro across the tropics along with the sun's zenith. Waller (1976) has pointed out that in the tropics the planting season of most crops coincides with the start of the rainy season, so that the cropping season moves to and fro in keeping with the movement of the intertropical convergence zone. The prevaling winds blow toward this moving zone of low pressure; they carry airborne spores from old crops to young as a natural consequence of the zone's movement, and so maintain a continuity of disease. Waller considers it to be highly probable that many of the common foliage pathogens of tropical crops, such as the rust fungi and fungi imperfecti with airborne spores, especially *Alternaria*, *Drechslera*, and *Cercospora*, are carried in this way from old to young crops after the progression of the rainy season across the tropics. The essential condition of endemic disease, that it be always present, is therefore well met within the zone as a whole. Apart from the intertropical convergence zone there are geographical factors such as proximity to oceans. Through their effect, parts of the tropics are wet with no well-defined dry season; parts have a bimodal distribution of rainfall with a double rainy season; and parts have a dry season when agriculture, if any, depends on irrigation.

It is in the wet tropics that disease, especially foliage disease, is most likely to be continuously present, that is, endemic, although even in the drier parts there is also a marked tilt towards overlapping seasons and endemicity. The same feature, continuous inoculum, that makes disease endemic is also likely to make diseases more numerous. In the tropics the number of diseases affecting any one crop seems to be much greater than in the temperate zone, as Renfro and Ullstrup (1973) noted for maize, Ou (1973) for rice, and Echandi (1973) for *Phaseolus* beans.

Because of the tilt towards endemic disease, there should be a corresponding tilt towards using horizontal resistance in the tropics. Theory predicts and experience confirms that against tropical endemic disease even a moderate increase of horizontal resistance can have a marked effect. Robinson (1979), among others, has stressed the importance of using horizontal resistance in the tropics.

16.7   APPENDIX

This appendix describes the model on which Table 16.1 was based. It is the simplest possible model of disease increase. It assumes that at any moment of time the rate of increase of disease is proportional both to the amount of infectious tissue, e.g., sporulating lesions, and to the amount of healthy susceptible tissue still available for infection.

The assumption is expressed by the equation

$$dx(t)/dt = \alpha[x(t - p) - x(t - 1 - p)][1 - x(t)] \qquad (16.1)$$

Here, α is the progeny/parent ratio. The infectious period is taken to be the unit of time; $x(t)$, $x(t - p)$, and $x(t - 1 - p)$ are x at times t, $t - p$, and $t - 1 - p$ respectively. Interpreted, $x(t)$ is the total proportion of diseased tissue, i.e., disease at all stages of development (latent, infectious, and removed) at current time. Similarly, $x(t - p)$ is the proportion of tissue that has been infected long enough to pass the latent period p and therefore is or was infectious; and $x(t - 1 - p)$ is the proportion of tissue that has been infected long enough to pass both the latent period p and the infectious period $i = 1$, and therefore was, but has ceased to be, infectious (in alternative terminology it is tissue that has been removed); $x(t - p) - x(t - 1 - p)$ is therefore the proportion of tissue actually infectious (e.g., sporulating) as distinct from infected (but not necessarily all sporulating); and $1 - x(t)$ is the proportion of healthy, susceptible tissue still unoccupied by the pathogen and available for infection. From Eq. (16.1) by appropriate integration, we derive Table 16.1.

Table 16.1 shows that at low progeny/parent ratios (i.e., in endemic disease) the pathogen cannot use all the susceptible tissue available for infection, even when given unlimited time to do so and even when it starts from a high level of initial inoculum. The reason is that endemic disease runs out of inoculum, i.e., of infectious as distinct from infected tissue. Tables 10.2, 10.3, 10.4, and 10.5 in Vanderplank (1982) illustrate this by numerical integration. Only at higher progeny/parent ratios above 5 (i.e., in epidemic disease) can the pathogen use all the susceptible tissue, and, given time, it can do so even when it starts from a very low level of initial inoculum.

Observe the contrast. Epidemic disease, given unlimited time, stops because it runs out of susceptible tissue to occupy, even though there is plenty of inoculum about. Endemic disease stops because it runs out of inoculum, even though there is still plenty of susceptible tissue left to occupy.*

The simple model involves some oversimplification. It assumes a random

---

* Ecologists err in assuming that K-strategists can use all the carrying capacity K of the habitat. MacArthur and Wilson (1967) define K-strategists by the logistic equation, $dN/dt = Nr(1 - N/K)$. The error arises from giving N two entirely different meanings within the one equation. The factor N in the equation is N (reproducing), the number of reproducing age in the population. Elsewhere in the equation N is N (total), the total number of all ages in the population. The ratio N (reproducing)/N (total) varies with N (with either meaning). Depending on their progeny/parent ratios, K-strategists use a highly variable proportion of K. The concept of a given carrying capacity K of the habitat is illusory because the habitat can never carry the number of K-strategists it has the resources for. There is a contradiction in ecological thought, and plant pathologists are advised to steer their own course.

distribution of disease. On this assumption multiple infection is allowed for by transforming the proportion of disease x into $-\ln(1 - x)$. The transform $-\ln(1 - x)$ is usually derived from the Poisson relation (Vanderplank, 1975); in our model it appears as the integral of $dx/(1 - x)$. Disease is seldom randomly distributed over a field, from field to field, or from bottom to top of the foliage. As a result the transform undercorrects. Whereas with random distribution a progeny/parent ratio of about 5 can be taken to indicate the border between endemic and epidemic disease, with nonrandom distribution the ratio would be higher. However, adjustment of the borderline does not alter the wider generalizations reached in this chapter, and we need not here pursue the problem of nonrandomness further.

Another simplification is in the concept of initial inoculum. This concept is realistic enough for epidemic disease, but in endemic disease both host and pathogen are renewed at intervals. The model applies to only one interval at a time, and a more general model has yet to be attempted.

17

Heterogeneous
Host Populations and
the Accumulation of
Resistance Genes

17.1 INTRODUCTION

Mixed varieties and multilines have been used in order to facilitate and speed the accumulation of resistance genes within the host population. There is no evidence that heterogeneity as such stabilizes resistance.

Advocates of heterogeneity appeal to nature. Wild plant populations, they argue, are heterogeneous, and this heterogeneity stabilizes them and protects them from disease. Modern agriculture increasingly demands uniform crops, and this, they argue, is the source of epidemics sweeping unchecked through a homogeneous substrate. The advocates of a return to nature seldom refer to climax plant communities dominated by a single species; and if they do they assume, without evidence, that the species must be sufficiently heterozygous to make for effective heterogeneity.

The resistance involved in the argument for heterogeneity is vertical. By definition, vertical resistance needs at least two host genotypes to exist. Heterogeneity within a field can provide the two or more genotypes and give the host population a resistance it otherwise might lack. The vertical resistance of the field can be greater than the resistance of any of the constituent genotypes. By definition, horizontal resistance is effective even in one host genotype. The horizontal resistance of a heterogeneous field will be intermediate, less than that of the most resistant constituent genotype and more than that of the least resistant.

There are two sorts of host heterogeneity. In the first sort, there is a mixture of taxa of relatively wide differences: species, genera, families, etc. Pastures of grass and clover or forests of maple and oak are examples. Species within the mixture are likely to be nonhosts to pathogens of other species. In the second sort, heterogeneity is given by mixtures of lines or varieties within a species or group of species that constitute an agricultural crop. It is this second sort of crop heterogeneity that concerns us.

17.2   MIXED VARIETIES AND MULTILINES

Mixed varieties are produced by the mechanical mixing of seed of existing cultivars. They have a long history. Rosen (1949), during the epidemic of Victoria blight in oats, suggested using an oat mixture of which about 70% were derivatives of the variety Victoria, which was susceptible to *Helminthosporium victoriae* but resistant to the then-prevalent population of *Puccinia coronata*, and 30% were derivatives of Bond, which was resistant to *H. victoriae* but susceptible to *P. coronata*. Recently, variety mixtures have been recommended for the control of barley powdery mildew caused by *Erysiphe graminis hordei*. Wolfe (1978) reported results of trials with a mixture of three barley cultivars, Hassan, Wing, and Midas. The mixture had somewhat less mildew than the mean of the three cultivars in pure stands.

A multiline or multiline cultivar is a set of about 10 lines, more or less, that are as near as possible isogenic except that each of the lines carries a different gene for resistance. The gene is introduced by backcrossing, and the multiline is prepared by mechanically mixing seed of each line. Borlang (1953, 1965) suggested wheat multilines for the control of stem rust caused by *P. graminis tritici*. Oat multilines have been extensively used in Iowa and neighboring states for the control of crown rust of oats caused by *P. coronata*.

Multilines have been studied by, among others, Cournoyer *et al.* (1968), Leonard and Kent (1968), Leonard (1969), Browning and Frey (1969), Frey *et al.* (1977), and Luthra and Rao (1979). The results of Luthra and Rao, working with wheat multilines for the control of leaf rust caused by *P. recondita*, are typical. A 5-line multiline had about one-third less leaf rust than the average of its component isolines; a 12-line multiline had about one-half less; and the best multiline, with 7 component isolines, had about two-thirds less.

Multilines of the International Maize and Wheat Improvement Center (CIMMYT) differ widely from traditional multilines just described. They are used by CIMMYT in a crash program designed primarily to protect wheat against the three rust fungi, *P. graminis*, *P. recondita*, and *P. striiformis*, although other pathogens and other crops are included. The program has been described by Dubin and Rajaram (1982). The CIMMYT multilines differ markedly, in that backcrossing has been discarded in favor of double, or four-way, crosses. This allows germ plasm from four parents to be combined in a single line. Double crosses allow breeders to combine many genes for resistance to one disease at one time and also enable them to combine genes against several diseases rapidly. The process is very efficient, and the components of one wheat multiline already have at least 16 distinct leaf rust resistance genes.

17.3 HETEROGENEITY VERSUS GENE ACCUMULATION

Many reports claiming the superiority of variety mixtures and multilines make unfair or irrelevant comparisons. Variety mixtures are compared with single varieties, multilines with single isolines. Plots or fields with many resistance genes are compared with plots or fields with few resistance genes. Any superiority is due to the many instead of the few genes, i.e., to the accumulation of genes. Heterogeneity per se does not enter the comparisons.

Consider the matter at its simplest, and suppose that three resistance genes *R1*, *R2*, and *R3* are available. A three-line multiline, we suppose, is prepared, each line having one of these genes. It is unfair and irrelevant to compare this multiline with a single line containing gene *R1*, *R2*, or *R3*. The relevant question is, would the multiline be better than a variety that combined all three genes? Would heterogeneity, i.e., dispersal of genes among the lines, be safer than uniformity, i.e., than accumulation of the same genes within a single variety? On all available information, the uniform single variety would be safer than the heterogeneous multiline. The uniform variety would be susceptible to race (1, 2, 3); so would every line in the multiline, so that against race (1, 2, 3) all relevant heterogeneity would disappear. The uniform variety would be resistant to races (1), (2), (3), (1, 2), (1, 3), and (2, 3); these races could all establish themselves on the multiline. The multiline would facilitate stepwise mutation to greater virulence; the uniform variety would not.

Heterogeneity per se does not increase resistance or make for greater safety. Increased resistance and greater safety come from the accumulation of resistance genes, be they dispersed or concentrated. The real aim of mixed varieties and multilines is gene accumulation, not heterogeneity.*

17.4   ESSENTIAL PURPOSE OF MIXED VARIETIES AND MULTILINES

Mixed varieties and multilines are shortcuts to accumulated resistance genes. They must be judged primarily as shortcuts. They obviate the slow process of accumulating genes within a single variety.

Mixing barley cultivars, as described by Wolfe (1978), makes sense in the battle against powdery mildew. It makes instant use of resistance genes already accumulated by barley breeders. The method has a limitation. Variety mixtures require a background of successful plant breeding; the varieties must preexist. It is doubtful whether plant breeders would develop new varieties simply for the sake of having them mixed.

Traditional multilines, compared with CIMMYT multilines, are slow to prepare. They have the added disadvantage that, unlike mixed varieties and CIMMYT multilines, they do not act against nontarget diseases. (Nontarget diseases are, e.g., oat diseases other than crown rust, if crown rust is the target of the multilines.) Their considerable background genetic uniformity is, however, an advantage if the crop has a fastidious market demanding uniformity in the product.

CIMMYT has the task of feeding a hungry world in a hurry, and CIMMYT multilines are well tailored for that purpose. No other known procedure could have accumulated 16 leaf rust resistance genes in a wheat population

---

* Mixed varieties and multilines should be seen in the light of an overriding axiom: Equal genetic resources provide greater and probably stabler resistance when concentrated than when dispersed. Put more elaborately, the axiom reads: A given set of resistance genes provides greater and probably stabler resistance when combined within a uniform population of plants (i.e., within a conventional variety) than when dispersed through a heterogeneous population (as in mixed varieties or multilines). By a given set is meant a given number of given genes. In the literature there is much exaggeration about the virtue of heterogeneity, which is confused with the abundance of the genetic resource. Uniformity has been blamed for the erosion of genetic resources. This axiom is about how to use equal genetic resources, contrasting genes accumulated within one variety with equal genes dispersed through a mixture or multiline. Mixed cropping, as when grass and clover are grown in mixture, is lauded; but (because grass is resistant to most clover diseases, and clover to most grass diseases) the genetic resources of a grass and clover mixture are much greater than those of grass alone or clover alone, and the axiom does not apply.

in so short a time. CIMMYT multilines do what multilines are meant to do : they put resistance genes into use fast. That this fast use involves heterogeneity is incidental and regrettable.

17.5   ROLE OF THE ABC–XYZ GROUPS: STABILIZING SELECTION VERSUS HETEROGENEITY

The ABC–XYZ system shows how greatly overrated the contribution of heterogeneity to safety can be. Return to Table 4.1, which gives information about the virulence of *P. graminis tritici* in the United States for 11 *Sr* genes in wheat and barley. The genes are *Sr6*, *Sr9a*, *Sr9b*, *Sr15*, and *Sr17* of the ABC group, and *Sr7b*, *Sr9e*, *Sr10*, *Sr11*, *SrTt1*, and *SrTmp* of the XYZ group. Suppose that an 11-line wheat multiline was grown, each of its 11 isolines having 1 of the *Sr* genes. In weather favorable for stem rust, the multiline would be severely attacked. More than 3/4 of the population of the stem rust fungus would be virulent on each of the lines with genes of the XYZ group, and a substantial proportion would be virulent on lines with genes of the ABC group. The multiline would be a failure. But uniform fields of a single conventional wheat variety that combined two of the genes, *Sr6* and *Sr9e*, would be resistant and escape stem rust all the way from Mexico to Canada (Roelfs and McVey, 1975). Other pairs, such as genes *Sr9b* and *SrTmp*, chosen from the same 11 genes and combined within a uniform variety would also be effective. The uniform varieties would be resistant because of stabilizing selection against virulence for the pairs of genes; that was the theme of Chapter 4. The stabilizing selection was no flash in the pan; it was consistent from year to year. It, and not heterogeneity, is the essential component of safe plant breeding.

CIMMYT insists that all isolines to be used in multilines should everywhere be free from rust when tested worldwide. It is to be hoped that this insistence automatically ensures enough stabilizing selection. Food for millions depends on it.

Bibliography

Adkisson, P. C. (1954). The influence of hybridity and boll load upon the incidence of Verticillium wilt in cotton. M.S. Thesis, Univ. of Arkansas, Fayètteville, Arkansas (quoted by Batson *et al.*, 1970).

Ahn, S. W., and Ou, S. H. (1982a). Quantitative resistance of rice to blast disease. *Phytopathology* **72,** 279–282.

Ahn, S. W., and Ou, S. H. (1982b). Epidemiological implications of the spectrum of resistance to rice blast. *Phytopathology* **82,** 282–284.

Alexander, L. J., and Hoover, M. M. (1955). Disease resistance in wild species of tomato. *Ohio Agric. Exp. Stn. Res. Bull.* **752.**

Alexander, L. J., and Tucker, C. M. (1945). Physiologic specialization in the tomato wilt fungus *Fusarium oxysporum* f. sp. *lycopersici. J. Agric. Res.* **70,** 303–313.

Allard, R. W. (1960). "Principles of Plant Breeding." Wiley, New York.

Allen, R. F. (1926). A cytological study of *Puccinia recondita* form 11 on Little Club wheat. *J. Agric. Res.* **33,** 201–222.

Anderson, M. G. (1982). Interpreting residual effects of "defeated" resistance genes. *Phytopathology* **72,** 1383–1384.

Anderson, R. L., and Wood, W. A. (1969). Carbohydrate metabolism in micro-organisms. *Annu. Rev. Microbiol.* **23,** 539–578.

Anikster, Y., and Wahl, I. (1979). Coevolution of the rust fungi on Gramineae and Liliaceae and their hosts. *Annu. Rev. Phytopathol.* **17,** 367–403.

Anonymous (1937). A study of the effect of leaf rust (*Puccinia triticina* Eriks.) on yield, composition, and quality of wheat. *Indiana Agric. Exp. Stn. Rep. 1936,* 9.

Anonymous (1954). Verslag van de enquete over het optreden van de aartappelziekte, *Phytophthora infestans* (Mont.) de Bary in 1953. *Jaarb. Plantenziektenkundige Dienst Wageningen,* pp. 34–53.

Anonymous (1958). West African Maize Research Institute. Review of research for the period January 1955–December 1957. Published by the Federal Department of Research, Ibadan, Nigeria.

Asai, G. N. (1960). Intra- and interregional movement of uredospores of black stem rust in the Upper Mississippi River Valley. *Phytopathology* **50,** 535–541.

Athwal, D. S., and Watson, I. A. (1957). Inheritance of resistance to wheat leaf rust in Mentana, a variety of *Triticum vulgare. Proc. Linn. Soc. N. S. W.* **82,** 245–252.

Batson, W. E., Bird, L. S., Tolmsoff, W. J., and Cater, C. M. (1970). Accumulation of gossypol and gossypol-like pigments associated with defruited cotton plants. *Phytopathology* **60,** 913–914.

Bell, A. A. (1973). Nature of disease resistance. *U.S. Dep. Agric., Agric. Res. Serv. ARS-S* **19,** 47–62.

Biffen, R. H. (1905). Mendel's laws of inheritance and wheat breeding. *J. Agric. Sci.* **1,** 4–48.

Bird, L. S. (1980). Breeding for disease and nematode resistance in cotton. *Proc. Intern. Short Course Host Plant Resistance, Texas A & M Univ., College Station, Texas,* pp. 86–100.

Black, W. (1960). Races of *Phytophthora infestans* and resistance problems in potatoes. *Scot. Plant Breeding Stn. Annu. Rep.,* pp. 29–38.

Black, W., Mastenbroek, C., Mills, W. R., and Petersen, L. C. (1953). A proposal for an international nomenclature of races of *Phytophthora infestans* and of genes controlling immunity in *Solanum demissum* derivatives. *Euphytica* **2,** 173–178.

Borlaug, N. E. (1946). *Puccinia sorghi* on corn in Mexico. *Phytopathology* **36,** 395.

Borlaug, N. E. (1953). New approach to the breeding of wheat varieties resistant to *Puccinia graminis tritici. Phytopathology* **43,** 467.

Borlaug, N. E. (1965). Wheat, rust, and people. *Phytopathology* **55,** 1088–1098.

Brinkerhoff, L. A. (1963). Variation of *Xanthomonas malvacearum,* the cotton blight organism. *Okla. Agric. Exp. Stn. Tech. Bull.* **T98,** 1–96.

Brinkerhoff, L. A. (1970). Variation in *Xanthomonas malvacearum* and its relation to control. *Annu. Rev. Phytopathol.* **8,** 85–110.

Browder, L. E. (1980). A compendium of information about named genes for low reaction to *Puccinia recondita* in wheat. *Crop Sci.* **20,** 775–779.

Browning, J. A., and Frey, K. J. (1969). Multiline cultivars as a means of disease control. *Annu. Rev. Phytopathol.* **7,** 355–362.

Browning, J. A., Cournoyer, B. M., Jowett, D., and Mellon, J. (1970). Urediospores and grain yields from interacting crown rust races and commercial multiline cultivars. *Phytopathology* **60,** 1286.

Caldwell, R. M., Roberts, J. J., and Eyal, Z. (1970). General resistance ("slow rusting") to *Puccinia recondita* f. sp. *tritici* in winter and spring wheats. *Phytopathology* **60,** 1287.

Cammack, R. H. (1961). *Puccinia polysora:* a review of some factors affecting the epiphytotic in West Africa. *Rep. 6th Commonwealth Mycol. Conf. 1960,* pp. 134–138.

Carson, M. L., and Hooker, A. L. (1982). Reciprocal translocation testcross analysis of genes for anthracnose stalk rot resistance in a corn inbred line. *Phytopathology* **72,** 175–177.

Caten, C. E. (1970). Spontaneous variability of single isolates of *Phytophthora infestans* II. Pathogenic variation. *Can. J. Bot.* **48,** 897–905.

Catherall, P. C., Jones, A. T., and Hayes, J. D. (1970). Inheritance and effectiveness of genes of barley that condition tolerance to barley yellow dwarf virus. *Ann. Appl. Biol.* **65,** 153–161.

Chester, K. S. (1946). "The Cereal Rusts." Chronica Botanica, Waltham, Massachusetts.

Cirulli, M., and Alexander, L. J. (1966). A comparison of pathogenic races of *Fusarium oxysporum* f. *lycopersici* and different sources of resistance in tomato. *Phytopathology* **56,** 1301–1304.

Cirulli, M., and Ciccarese, F. (1975). Interactions between TMV isolates, temperature, allelic condition and combination of the *Tm* resistance genes in tomato. *Phytopathol. Mediterr.* **14,** 100–105.

Clarke, A. E., Jones, H. A., and Little, T. M. (1944). Inheritance of bulb color in the onion. *Genetics* **29,** 569–575.

Clifford, B. C. (1972). The histology of race non-specific resistance to *Puccinia hordei* Orth. in barley. *Proc. Europ. Mediterr. Cereal Rusts Conf. Prague* **1,** 75–79.

Clifford, B. C. (1975). Stable resistance to cereal disease: problems and progress. *Rep. Welsh Plant Breed. Stn. 1974,* pp. 107–113.

Clifford, B. C., and Clothier, R. B. (1974). Physiologic specialization of *Puccinia hordei* on barley hosts with non-hypersensitive resistance. *Trans. Br. Mycol. Soc.* **63,** 421–430.

Cornu, M. (1881). Prolongation de l'activité végétale des cellules chlorophylliennes sous l'influence d'un parasite. *C.R. Hebd. Séances Acad. Sci. Paris* **93,** 1162–1164.

Cournoyer, B. M., Browning, J. A., and Jowett, D. (1968). Crown rust intensification within and dissemination from pure line and multiline varieties of oats. *Phytopathology* **58,** 1047.

Cox, A. E., and Large, E. C. (1960). Potato blight epidemics throughout the world. *U.S. Dep. Agric., Agric. Handb.* **174.**

Craig, J., and Hooker, A. L. (1961). Relation of sugar trends and pith density to Diplodia stalk rot in dent corn. *Phytopathology* **51,** 376–382.

Crill, P., Jones, J. P., Burgis, D. S., and Woltz, S. S. (1972). Controlling *Fusarium* wilt of tomato with resistant varieties. *Plant. Dis. Rep.* **56,** 695–699.

Cross, J. E. (1963). Pathogenic differences in Tanganyika populations of *Xanthomonas malvacearum. Emp. Cotton Grow. Rev.* **40,** 125–130.

Crosse, J. E. (1975). Variation amongst plant pathogenic bacteria. *Ann. Appl. Biol.* **81,** 834.

Day, P. R. (1974). "Genetics of Host–Parasite Interaction." Freeman, San Francisco, California.

de Bary, A. (1876). Researches into the nature of the potato fungus (*Phytophthora infestans*). *J. R. Agric. Soc. Engl.* **12,** 239.

de Bruijn, H. L. G. (1951). Pathogenic differentiation in *Phytophthora infestans* (Mont.) de Bary. *Phytopathol. Z.* **18,** 339–359.

Delp, C. J. (1954). Effect of temperature and humidity on the grape powdery mildew fungus. *Phytopathology* **44,** 615–626.

Denward, T. (1970). Differentiation in *Phytophthora infestans.* II. Somatic recombination in vegetative mycelium. *Hereditas* **66,** 35–48.

De Turk, E. E., Earley, E. B., and Holbert, J. R. (1937). Resistance of corn hybrids related to carbohydrates. *Ill. Agr. Exp. Stn. Annu. Rep.* **49,** 43–45.

Dodd, J. L. (1980a). The role of plant stresses in the development of corn stalk rots. *Plant Dis.* **64,** 533–537.

Dodd, J. L. (1980b). Grain sink size and predisposition of *Zea mays* to stalk rot. *Phytopathology* **70,** 534–535.

Dubin, H. J., and Rajaram, S. (1982). The CIMMYT's international approach to breeding disease-resistant wheat. *Plant. Dis.* **66,** 967–971.

Durrell, L. W., and Parker, J. H. (1920). Comparative resistance of varieties of oats to crown and stem rusts. *Iowa Agric. Exp. Stn. Res. Bull.* **62.**

Dyck, P. L., and Samborski, D. J. (1968). Genetics of resistance to leaf rust in the common wheat varieties Webster, Loris, Brevit, Carina, Malakof, and Centenario. *Can. J. Genet. Cytol.* **10,** 7–17.

Dyck, P. L., Samborski, D. J., and Anderson, R. G. (1966). Inheritance of adult-plant leaf rust resistance derived from the common wheat varieties Exchange and Frontana. *Can. J. Genet. Cytol.* **8,** 665–671.

Eaton, F. M., and Ergle, D. R. (1953). Relationship of seasonal trends in carbohydrate and nitrogen levels and effects of spraying with sucrose and urea to nutritional interpretations of boll shedding in cotton. *Plant Physiol.* **28,** 503–519.

Eaton, F. M., and Rigler, N. E. (1946). Influence of carbohydrate levels and root-surface microfloras on Phymatotrichum root rot in cotton and maize plants. *J. Agric. Res.* **72,** 137–161.

Echandi, E. (1973). The pathology of bean (*Phaseolus vulgaris*) grown under tropical and temperate climates. *Int. Congr. Plant Pathol., 2nd, Minneapolis,* Abst. 1045.

Eide, C. J., Bonde, R., Gallegly, M. E., Graham, K. M., Mills, W. R., Niederhauser, J., and Wallin, J. R. (1959). Report of the late blight investigations committee. *Am. Potato J.* **36,** 421–423.

Elliott, C., and Jenkins, M. T. (1946). Helminthosporium leaf blight of corn. *Phytopathology* **36,** 660–666.

Fajemisin, J. M., and Hooker, A. L. (1974). Predisposition to Diplodia stalk rot in corn affected by three Helminthosporium leaf blights. *Phytopathology* **64,** 1496–1499.

Finlay, K. W. (1952). Inheritance of spotted wilt resistance in the tomato. I. Identification of strains of the virus by the resistance or susceptibility of tomato species. *Aust. J. Sci. Res. Ser.B* **5,** 303–314.

Fisher, R. A. (1930). "The Genetical Theory of Natural Selection." Oxford Univ. Press, (Clarendon), London and New York.

Flor, H. H. (1942). Inheritance of pathogenicity of *Melampsora lini. Phytopathology* **32,** 653–669.

Flor, H. H. (1953). Epidemiology of flax rust in the North Central States. *Phytopathology* **43,** 624–628.

Flor, H. H. (1956). The complementary genic systems in flax and flax rust. *Adv. Genet.* **8,** 29–54.

Flor, H. H. (1958). Mutation to wider virulence in *Melampsora lini. Phytopathology* **48,** 297–301.

Flor, H. H. (1960). The inheritance of X-ray mutations to virulence in a uredospore culture of race 1 of *Melampsora lini. Phytopathology* **50,** 603–605.

Flor, H. H. (1971). Current status of the gene-for-gene concept. *Annu. Rev. Phytopathol.* **9,** 275–296.

Ford, E. B. (1945). Polymorphisms. *Biol. Rev.* **20,** 73–88.

Frandsen, N. V. (1956). Rasse 4 von *Phytophthora infestans* in Deutschland. *Phytopathol. Z.* **26,** 124–130.

Frederiksen, R. A., and Rosenow, D. T. (1980). Breeding for disease resistance in sorghum. *Proc. Int. Short Course Host Plant Resistance, Texas A & M Univ., College Station, Texas,* pp. 137–167.

Frey, K. J., Browning, J. A., and Simons, M. D. (1977). Management systems for host genes to control disease loss. *Ann. N.Y. Acad. Sci.* **287,** 255–274.

Fromme, F. D. (1913). The culture of rusts in the greenhouse. *Bull. Torrey Bot. Club* **40,** 501–521.

Gassner, G., and Kirchhoff, H. (1934). Einige vergleichende Versuche über Verschiebungen der Rostresistenz in Afhänzigkeit vom Entwicklungszustand der Getreidepflanzen. *Phytopathol. Z.* **7,** 43–52.

Gassner, G., and Straib, W. (1932). Über mutationen in eines biologischen Rasse von *Puccinia glumarum* (Schmidt) Erikss. und Henn. *Z. Indukt. Abstamm. Vererbungsl.* **43,** 155–180.

Gavinlertvatana, S., and Wilcoxson, R. D. (1978). Inheritance of slow rusting of spring wheat by *Puccinia recondita* f. sp. *tritici* and host parasite relationships. *Trans. Br. Mycol. Soc.* **71,** 413–418.

Gerdemann, J. W., and Finley, A. M. (1951). The pathogenicity of races 1 and 2 of *Fusarium oxysporum* f. sp. *lycopersici. Phytopathology* **41,** 238–244.

Gibbs, A. F., and Wilcoxson, R. D. (1972). Effect of sugar content of *Poa pratensis* on Helmin-thosporium leaf spot. *Physiol. Plant Pathol.* **2**, 279–287.

Gill, C. C., and Buchannon, K. W. (1972). Reaction of barley hybrids from crosses with C.I. 5791 to four isolates of barley yellow dwarf virus. *Can. J. Plant Sci.* **52**, 305–309.

Graham, K. M., Dionne, L. A., and Hodgson, W. A. (1961). Mobility of *Phytophthora infestans* on blight resistant selections of potatoes and tomatoes. *Phytopathology* **51**, 264–265.

Grainger, J. (1956). Host nutrition and attack by fungal parasites. *Phytopathology* **46**, 445–456.

Grainger, J. (1968). Cp/Rs and the disease potential of plants. *Hortic. Res.* **8**, 1–40.

Green, G. J. (1971). Stem rust of wheat, barley, and rye in Canada in 1970. *Can. Plant Dis. Surv.* **51**, 20–23.

Green, G. J. (1972a). Stem rust of wheat, barley, and rye in Canada in 1971. *Can. Plant Dis. Surv.* **52**, 11–14.

Green, G. J. (1972b). Stem rust of wheat, barley, and rye in Canada in 1972. *Can. Plant Dis. Surv.* **52**, 162–167.

Green, G. J. (1974). Stem rust of wheat, barley, and rye in Canada in 1973. *Can. Plant Dis. Surv.* **54**, 11–16.

Green, G. J. (1975). Stem rust of wheat, barley, and rye in Canada in 1974. *Can. Plant Dis. Surv.* **55**, 51–57.

Green, G. J. (1976a). Stem rust of wheat, barley, and rye in Canada in 1975. *Can. Plant Dis. Surv.* **56**, 15–18.

Green, G. J. (1976b). Stem rust of wheat, barley, and rye in Canada in 1976. *Can. Plant Dis. Surv.* **56**, 119–122.

Green, G. J. (1978). Stem rust of wheat, barley, and rye in Canada in 1977. *Can. Plant Dis. Surv.* **58**, 44–48.

Green, G. J. (1979). Stem rust of wheat, barley, and rye in Canada in 1978. *Can. Plant Dis. Surv.* **59**, 43–47.

Green, G. J. (1980). Physiologic races and epidemiology of *Puccinia graminis* on wheat, barley, and rye in Canada in 1979. *Can. J. Plant Pathol.* **2**, 241–245.

Green, G. J. (1981). Identification of physiologic races of *Puccinia graminis* f. sp. *tritici* in Canada. *Can. J. Plant Pathol.* **3**, 33–39.

Green, G. J., and Campbell, A. B. (1979). Wheat cultivars resistant to *Puccinia graminis tritici* in Western Canada: their development, performance, and economic value. *Can. J. Plant Pathol.* **1**, 3–11.

Gregory, L. V., Ayers, J. E., and Nelson, R. R. (1981). Reliability of the apparent infection rate in epidemiological research. *Phytopathol. Z.* **100**, 135–142.

Grogan, C. D., and Rosenkranz, E. E. (1968). Genetics of host reaction to corn stunt virus. *Crop Sci.* **8**, 251–254.

Hanada, K., and Harrison, B. D. (1977). Effects of virus genotype and temperature on seed transmission of nepoviruses. *Ann. Appl. Biol.* **85**, 79–82.

Harlan, J. R. (1961). Geographic origin of plants useful in agriculture. *In* "Germ Plasm Resources" (R. E. Hodgson, ed.), pp. 3–19, AAAS Publication 66.

Harlan, J. R. (1972). Genetics of disaster. *J. Environ. Qual.* **1**, 212–215.

Harlan, J. R. (1980). Genetics of agriculture and crop evolution. *Proc. Int. Short Course Host Plant Resistance, Texas A & M Univ., College Station, Texas*, pp. 1–8.

Heagle, A. S., and Moore, M. B. (1970). Some effects of adult resistance to crown rust of oats. *Phytopathology* **60**, 461–466.

Hennig, B., and Wittmann, H. G. (1972). Tobacco mosaic virus: mutants and strains. *In* "Principles and Techniques in Plant Virology" (C. I. Kado and H. O. Agrawal, eds.), pp. 546–594. Van Nostrand-Reinhold, Princeton, New Jersey.

Hodgson, W. A. (1961). Laboratory testing of the potato for partial resistance to *Phytophthora infestans. Am. Potato J.* **38,** 259–264.

Hogen Esch, J. A., and Zingstra, H. (1957). "Geniteurslijst voor aartappelrassen 1957." Commissie ter bevordering van het kweken en het onderzoek van nieuwe aartappelrassen, Wageningen, Netherlands.

Holbert, J. R., Hoppe, P. E., and Smith, A. L. (1935). Some factors affecting infection with and spread of *Diplodia zeae* in host tissue. *Phytopathology* **25,** 1113–1114.

Hooker, A. L. (1962). Corn leaf disease. *Proc. 17th Annu. Hybrid Corn Indust. Res. Conf.* **17,** 24–36.

Hooker, A. L. (1967a). The genetics and expression of resistance in plants to rusts of the genus *Puccinia. Annu. Rev. Phytopathol.* **5,** 163–182.

Hooker, A. L. (1967b). Inheritance of mature plant resistance to rust in corn. *Phytopathology* **57,** 815.

Hooker, A. L. (1979). Breeding for resistance to some complex disease of corn. *In* "International Rice Research Institute Rice Blast Workshop," pp. 153–181. Los Banos, Laguna, Philippines.

Hooker, A. L., and Le Roux, P. M. (1957). Sources of protoplasmic resistance to *Puccinia sorghi* in corn. *Phytopathology* **47,** 187–191.

Hooker, A. L., and Saxena, K. M. S. (1971). Genetics of disease resistance in plants. *Annu. Rev. Genet.* **5,** 407–421.

Horsfall, J. G. (1975). The story of a nonconformist. *Annu. Rev. Phytopathol.* **13,** 1–13.

Horsfall, J. G., and Cowling, E. B. (1978). Pathometry: the measurement of plant disease. *In* "Plant Disease" (J. G. Horsfall and E. B. Cowling, eds.), Vol. II, pp. 119–137. Academic Press, New York.

Horsfall, J. G., and Dimond, A. E. (1957). Interactions of tissue sugar, growth substances, and disease susceptibility. *Z. Pflanzenkr.* **64,** 415–421.

Horsfall, J. G., and Heuberger, J. W. (1942). Causes, effects, and control of defoliation on tomatoes. *Conn. Agric. Exp. Stn. Bull., New Haven* **456,** 182–223.

Howard, H. W., Johnson, R., Russell, G. E., and Wolfe, M. S. (1970). Problems in breeding for resistance to diseases and pests. *Rep. Plant Breeding Inst. Cambridge 1969,* pp. 6–36.

Howatt, J. L., and Grainger, P. N. (1955). Some new findings concerning *Phytophthora infestans* (Mont.) de By. *Am. Potato J.* **32,** 180–188.

Hoyle, M. C. (1977). High resolution of peroxidase-indoleacetic acid oxidase enzymes from horseradish by isoelectric focussing. *Plant Physiol.* **60,** 787–793.

Hughes, G. R., and Hooker, A. L. (1971). Genes conditioning resistance to northern leaf blight in maize. *Crop Sci.* **11,** 180–184.

Jenkins, M. T., and Robert, A. L. (1961). Further genetic studies of resistance to *Helminthosporium turcicum* Pass. in maize by means of chromosomal translocations. *Crop Sci.* **1,** 450–455.

Johnson, D. A., and Wilcoxson, R. D. (1978). Components of slow-rusting in barley infected with *Puccinia hordei. Phytopathology* **68,** 1470–1474.

Johnson, R., and Law, C. N. (1975). Genetic control of durable resistance to yellow rust (*Puccinia striiformis*) in the wheat cultivar Hybride de Bersée. *Ann. Appl. Biol.* **81,** 385–391.

Johnston, C. O., and Browder, L. E. (1966). Seventh revision of the international register of physiologic races of *Puccinia recondita* f. sp. *tritici. Plant Dis. Rep.* **50,** 576.

Kammermann, N. (1950). Undersökninger rörande Potatisbladmöglet, *Phytophthora infestans* (Mont.) de By. I. Metodologiskundersökning angaende prövningen av potatisblastens resistens mot bladmöglet. *Statens Medd. Vaextsdyddsanst. Stockholm* **57.**

Katan, J., and Ausher, R. (1974). Distribution of race 2 of *Fusarium oxysporum* f. sp. *lycopersici* in tomato fields in Israel. *Phytoparasitica* **2**, 83–90.

Katan, J., and Wahl, I. (1969). Occurrence in Israel of new dangerous isolates of the Fusarium wilt pathogen. *Proc. 1st Congr. Mediterr. Phytopathol. Union*, pp. 425–430.

Katsuya, K., and Green, G. J. (1967). Reproductive potentials of races 15B and 56 of wheat stem rust. *Can. J. Bot.* **45**, 1077–1091.

Kim, S. K., and Brewbaker, J. L. (1977). Inheritance of general resistance in maize to *Puccinia sorghi*. *Crop Sci.* **17**, 456–461.

Kirste (1958). Ergebnisse von Krautfäule-Spritzversuchen. *Kartoffelbau* **9**, 114–115.

Knott, D. R., and Dvořák, J. (1976). Alien germ plasm as a source of resistance to disease. *Annu. Rev. Phytopathol.* **14**, 211–235.

Knutson, K. W., and Eide, C. J. (1961). Parasitic aggressiveness in *Phytophthora infestans*. *Phytopathology* **51**, 286–290.

Kochman, J. K., and Brown, J. F. (1975). Development of the stem and crown rust fungi on leaves, sheaths, and peduncles of oats. *Phytopathology* **65**, 1404–1408.

Koehler, B. (1960). Cornstalk rots in Illinois. *Univ. Ill. Agric. Exp. Stn. Bull. 658.*

Krull, C. F., Reyes, R., Orjuela, J., and Bustamante, E. (1965). Importance of "small-uredia" reaction as an index of partial resistance to oat stem rust in Colombia. *Crop. Sci.* **5**, 494–497.

Kuhn, R. C., Ohm, H. W., and Shaner, G. E. (1978). Slow leaf-rusting resistance in wheat against twenty two isolates of *Puccinia recondita*. *Phytopathology* **68**, 651–656.

Kulkarni, R. N., and Chopra, V. L. (1982). Environment as the cause of differential interaction between host cultures and pathogenic races. *Phytopathology* **72**, 1384–1386.

Lapwood, D. H. (1961). Potato haulm resistance to *Phytophthora infestans*. II. Lesion production and sporulation. *Ann. Appl. Biol.* **49**, 316–330.

Latin, R. X., MacKenzie, D. R., and Cole, H. (1981). The influence of host and pathogen genotypes on the apparent infection rates of potato late blight epidemics. *Phytopathology* **71**, 82–85.

Leijerstam, B. (1972). Studies in powdery mildew of wheat in Sweden. III. Variability of virulence in *Erysiphe graminis* f. sp. *tritici* due to gene recombination and mutation. *Natl. Swed. Inst. Plant Prot., Contrib.* **15**, 231–248.

Leonard, K. J. (1969). Factors affecting rates of stem rust increase in mixed plantings of susceptible and resistant oat varieties. *Phytopathology* **59**, 845–850.

Leonard, K. J., and Kent, G. C. (1968). Increase of stem rust in mixed plantings of susceptible and resistant oat varieties. *Phytopathology* **58**, 400–401.

Leppik, E. E. (1966). Searching gene centers of the genus *Cucumis* through host-parasite relationships. *Euphytica* **15**, 323–335.

Leppik, E. E. (1970). Gene centers of plants as sources of disease resistance. *Annu. Rev. Phytopathol.* **8**, 323–344.

Lerner, L. M. (1954). "Genetic Homeostasis." Wiley, New York.

Lerner, L. M. (1958). "The Genetic Basis of Selection." Wiley, New York.

Loegering, W. Q., and Sears, E. R. (1981). Genetic control of disease expression in stem rust of wheat. *Phytopathology* **71**, 425–428.

Loomis, W. E. (1934). The translocation of carbohydrates in maize. *Iowa State Coll. J. Sci.* **9**, 509–520.

Luig, N. H. (1979). Mutation studies in *Puccinia graminis tritici*. *Proc. 5th Int. Wheat Genet. Symp.*, pp. 533–539.

Luig, N. H. (1980). Wheat rust survey 1979–80. *Rep. Univ. Sydney Plant Breeding Inst.*

Luig, N. H. (1981). Wheat rust survey 1980–81. *Rep. Univ. Sydney Plant Breeding Inst.*

Luig, N. H., and Baker, E. P. (1973). Variability in oat stem rust in eastern Australia. *Proc. Linn. Soc. N. S. W.* **98,** 53–61.

Luig, N. H., and Watson, I. A. (1965). Studies on the genetic nature of resistance to *Puccinia graminis* var. *tritici* in six varieties of common wheat. *Proc. Linn. Soc. N. S. W.* **90,** 299–327.

Luig, N. H., and Watson, I. A. (1970). The effect of complex genetic resistance in wheat on the variability of *Puccinia graminis* f. sp. *tritici. Proc. Linn. Soc. N. S. W.* **95,** 22–45.

Luke, H. H., Chapman, W. H., and Barnett, R. D. (1972). Horizontal resistance of Red Rustproof oats to crown rust. *Phytopathology* **62,** 414–417.

Luke, H. H., Barnett, R. D., and Chapman, W. H. (1975a). Types of horizontal resistance of oats to crown rust. *Plant Dis. Rep.* **59,** 332–334.

Luke, H. H., Barnett, R. D., and Pfahler, P. C. (1975b). Inheritance of horizontal resistance to crown rust in oats. *Phytopathology* **65,** 631–632.

Lukens, R. J. (1970). Melting-out of Kentucky bluegrass, a low sugar disease. *Phytopathology* **60,** 1276–1278.

Lupton, F. G. H., and Macer, R. C. F. (1962). Inheritance of resistance to yellow rust (*Puccinia glumarum* Erikss. and Henn.) in seven varieties of wheat. *Trans. Br. Mycol. Soc.* **45,** 21–45.

Luthra, J. K., and Rao, M. V. (1979). Multiline cultivars—how their resistance influences leaf rust disease in wheat. *Euphytica* **28,** 137–144.

MacArthur, R. H., and Wilson, E. O. (1967). "The Theory of Island Biogeography." Princeton Univ. Press, Princeton, New Jersey.

McIntosh, R. A. (1973). A catalogue of gene symbols for wheat. *Proc. 4th Int. Wheat Genet. Symp.,* pp. 893–937.

McIntosh, R. A. (1977). Nature of induced mutations affecting disease reactions in wheat. *In* "Induced Mutations against Plant Diseases," pp. 551–565. Int. At. Energy Ag., Vienna.

Magasanik, B. (1961). Catabolite repression. *Cold Spring Harbor Symp. Quant. Biol.* **26,** 249–254.

Mains, E. B. (1917). The relation of some rusts to the physiology of their hosts. *Am. J. Bot.* **4,** 179–220.

Malcolmson, J. F. (1969). Races of *Phytophthora infestans* occurring in Great Britain. *Trans. Br. Mycol. Soc.* **53,** 417–423.

Malcolmson, J. F. (1970). Vegetative hybridity in *Phytophthora infestans. Nature (London)* **225,** 971–972.

Martens, J. W., McKenzie, R. H., and Green, G. J. (1967). Thermal stability of stem rust resistance in oat seedlings. *Can. J. Bot.* **45,** 451–458.

Martin, C. D., Littlefield, L. J., and Miller, J. D. (1977). Development of *Puccinia graminis* f. sp. *tritici* in seedling plants of slow-rusting wheats. *Trans. Br. Mycol. Soc.* **68,** 161–166.

Martin, C. D., Miller, J. D., Busch, R. H., and Littlefield, L. J. (1979). Quantitation of slow rusting in seedlings and adult spring wheat. *Can. J. Bot.* **57,** 1550–1556.

Martin, T. J., and Ellingboe, A. H. (1976). Differences between compatible parasite/host genotypes involving the *Pm4* locus of wheat and the corresponding genes in *Erysiphe graminis* f. sp. *tritici. Phytopathology* **66,** 1435–1438.

Mather, K. (1943). Polygenic inheritance and natural selection. *Biol. Rev.* **18,** 32–64.

Mengden, K. (1975). Ultrastructural demonstration of different peroxidase activities during the bean rust infection process. *Physiol. Plant Pathol.* **6,** 275–282.

Mercado, A. C., and Lantican, R. M. (1961). The susceptibility of cytoplasmic male sterile lines of corn to *Helminthosporium maydis* Nish. and Miy. *Philipp. Agr.* **45,** 235–243.

Mercer, P. C., Wood, R. K. S., and Greenwood, A. D. (1974). Resistance to anthracnose of French bean. *Physiol. Plant Pathol.* **4,** 291–306.

Messiaen, C. M. (1957). Richesse en sucre des tiges de mais et verse parasitaire. *Rev. Pathol. Veg. Entomol. Agric. Fr.* **36**, 209–213.

Mortimore, C. G., and Ward, G. M. (1964). Root and stalk rot of corn in southwestern Ontario. III. Sugar levels as a measure of plant vigor and resistance. *Can. J. Plant Sci.* **44**, 451–457.

Müller, K. O. (1931). Über die Entwicklung von *Phytophthora infestans* auf anfälligen und widerstandsfähigen Kartoffelsorten. *Arb. Biol. Anst. Dahlem* **18**, 465–505.

Müller, K. O. (1953). The nature of resistance of the potato plant to blight. *Phytophthora infestans. J. Natl. Inst. Agr. Bot. (G. B.)* **6**, 346–360.

Murant, A. F., Taylor, C. E., and Chambers, J. (1968). Properties, relationships, and transmission of a strain of raspberry virus infecting raspberry cultivars immune to the common Scottish strain. *Ann. Appl. Biol.* **61**, 175–186.

Murray, M. J. (1969). Successful use of irradiation breeding to obtain *Verticillium*-resistant strains of peppermint, *Mentha piperita* L. *In* "Induced Mutations in Plants," pp. 345–371. Int. At. Energy Ag., Vienna.

Nass, H. A., Pedersen, W. L., MacKenzie, D. R., and Nelson, R. R. (1981). The residual effect of some "defeated" powdery mildew resistance genes in isolines of Chancellor winter wheat. *Phytopathology* **71**, 1315–1318.

Nelsen, R. R., Pedersen, W. L., and MacKenzie, D. R. (1982). The effect of pyramiding "defeated" wheat powdery mildew resistance genes on components of "slow mildewing." *Phytopathology* **72**, 932.

Newton, M., and Johnson, T. (1939). A mutation for pathogenicity in *Puccinia graminis tritici. Can. J. Res. Sect. C* **17**, 297–299.

Ohm, H. W., and Shaner, G. E. (1976). Three components of slow leaf-rusting at different growth stages in wheat. *Phytopathology* **66**, 1356–1360.

Ou, S. H. (1973). Contrasting pathological problems of rice under tropical and temperate climates. *Int. Congr. Plant Pathol., 2nd, Minneapolis,* Abst. 1046.

Paigen, K., and Williams, B. (1970). Catabolite repression and other control mechanisms in carbohydrate utilization. *Adv. Microb. Physiol.* **4**, 251–324.

Parlevliet, J. E. (1976). Partial resistance of barley to leaf rust, *Puccinia hordei.* III. The inheritance of the host plant effect on the latent period of four cultivars. *Euphytica* **25**, 241–248.

Patil, S. S., and Dimond, A. E. (1968). Repression of polygalacturonase synthesis in *Fusarium oxysporum* f. sp. *lycopersici* by sugars and its effect on symptom reduction in infected tomato plants. *Phytopathology* **58**, 676–682.

Paxman, G. J. (1963). Variation in *Phytophthora infestans. Europ. Potato J.* **6**, 14–23.

Pelham, J. (1972). Strain-genotype interactions of tobacco mosaic virus in tomato. *Ann. Appl. Biol.* **71**, 219–228.

Pelham, J., Fletcher, J. T., and Hawkins, J. H. (1970). The establishment of a new strain of tobacco mosaic virus resulting from the use of resistant varieties of tomato. *Ann. Appl. Biol.* **65**, 293–297.

Person, C., and Ebba, T. (1975). Genetics of fungal pathogens. *Genet. Supp.* **79**, 397–408.

Price, H. J. (1976). Evolution of DNA content in higher plants. *Bot. Rev.* **42**, 27–51.

Priestley, R. H. (1978). Detection of increased virulence in populations of wheat yellow rust. *In* "Plant Disease Epidemiology" (P. R. Scott and A. Bainbridge, eds.), pp. 63–70. Blackwell, Oxford.

Putter, C. A. J. (1980). The management of epidemic levels of endemic disease under tropical subsistence farming. *In* "Comparative Epidemiology" (J. Palti and J. Kranz, eds.), pp. 93–103. Center for Agricultural Publishing and Documentation, Wageningen, Netherlands.

Rao, Y. P., Mohan, S. K., and Reddy, P. R. (1971). Pathogenic variability in *Xanthomonas oryzae. Plant Dis. Rep.* **55**, 393–395.

Reddick, D., and Mills, W. R. (1938). Building up virulence in *Phytophthora infestans*. *Am. Potato J.* **15**, 29–34.

Reddy, O. R., and Ou, S. H. (1976). Pathogenic variability in *Xanthomonas oryzae*. *Phytopathology* **66**, 906–909.

Reed, G. M. (1914). Influence of light on infection of certain hosts by powdery mildew. *Science* **39**, 294–295.

Rees, R. B., Thompson, J. P., and Goward, E. A. (1979). Slow rusting and tolerance to rusts in wheat. II. The progress and effects of epidemics of *Puccinia recondita tritici* in selected wheat cultivars. *Aust. J. Agric. Res.* **30**, 421–432.

Reichert, I. (1958). Fungi and plant diseases in relation to biogeography. *Trans. N. Y. Acad. Sci.* **20**, 233–239.

Renfro, B. L., and Ullstrup, A. J. (1973). A comparison of maize diseases in temperate and tropical environments. *Int. Congr. Plant Pathol., 2nd, Minneapolis*, Abst. 1043.

Riley, R. (1973). Genetic changes in hosts and the significance of disease. *Ann. Appl. Biol.* **75**, 128–132.

Roane, C. W., Stakman, E. C., Loegering, W. Q., Stewart, D. M., and Watson, W. M. (1960). Survival of physiological races of *Puccinia graminis* var. *tritici* on wheat near barberry bushes. *Phytopathology* **50**, 40–44.

Robinson, P. W., and Hodges, C. F. (1977). Effect of nitrogen fertilizer on free amino acid and soluble sugar content of *Poa pratensis* and on infection and disease severity by *Drechslera sorokiniana*. *Phytopathology* **67**, 1239–1244.

Robinson, R. A. (1976). "Plant Pathosystems." Springer-Verlag, Berlin and New York.

Robinson, R. A. (1979). Permanent and impermanent resistance to crop parasites: A reexamination of the pathosystem concept with special reference to rice blast. *Z. Pflanzenzücht.* **83**, 1–39.

Roelfs, A. P., and Groth, J. V. (1980). A comparison of virulence phenotypes in wheat stem rust populations reproducing sexually and asexually. *Phytopathology* **70**, 855–862.

Roelfs, A. P., and McVey, D. V. (1975). Races of *Puccinia graminis* f. sp. *tritici* in the U.S.A. during 1974. *Plant Dis. Rep.* **59**, 681–685.

Rosen, H. R. (1949). Oat percentage and procedures for combining resistance to crown rust, including race 45, and Helminthosporium blight. *Phytopathology* **39**, 20.

Rouse, D. I., MacKenzie, D. R., and Nelson, R. R. (1981). A relationship between initial inoculum and apparent infection rate in a set of disease progress data for powdery mildew of wheat. *Phytopathol. Z.* **100**, 143–149.

Rowell, J. B. (1953). Leaf blight of tomato and potato plants. *R. I. Agric. Exp. Stn. Bull.* **320**, 1–29.

Rowell, J. B. (1981a). Relation of postpenetration events in Idaed 59 wheat seedlings to low receptivity to infection by *Puccinia graminis* f. sp. *tritici*. *Phytopathology* **71**, 732–736.

Rowell, J. B. (1981b). The relationship between slow rusting and a specific resistance gene for wheat stem rust. *Phytopathology* **71**, 1184–1186.

Rowell, J. B. (1982). Control of wheat stem rust by low receptivity to infection conditioned by a single dominant gene. *Phytopathology* **72**, 297–299.

Rowell, J. B., and McVey, D. V. (1974). Field evaluation of wheats of low receptivity to infection by stem rust. *Proc. Am. Phytopathol. Soc.* **1**, 89.

Rodorf, W. (1959). Problems of collection, maintenance and evaluation of wild species of cultivated plants. *FAO Plant Introd. Newsletter* **5**, 1–4.

Rudorf, W., and Schaper, P. (1951). Grundlagen und Ergebnisse der Züchtung kraukfäuleresistenter Kartoffelsorten. *Z. Pflanzenzücht.* **30**, 29–88.

Russell, G. E. (1976). Characterization of adult plant resistance to yellow rust in wheat. *Proc. 4th Europ. Mediterr. Cereal Rusts Conf.*, pp. 21–23.

Russell, W. A. (1965). Effect of corn leaf rust on grain yield and moisture in corn. *Crop Sci.* **5**, 95–96.

Samborski, D. J., and Dyck, P. L. (1982). Enhancement of resistance to *Puccinia recondita* by interaction of resistance genes in wheat. *Can. J. Plant Pathol.* **4**, 152–156.

Sands, D. C., and Lukens, R. L. (1974). Effect of glucose and adenosine phosphates on production of extracellular carbohydrates of *Alternaria solani*. *Plant Physiol.* **54**, 666–669.

Scandalios, J. G. (1974). Isozymes in development and differentiation. *Annu. Rev. Plant Physiol.* **25**, 225–258.

Schaper, P. (1951). Die Bedeutung der Inkubationzeit für die Züchtung krautfäuleresistenter Kartoffelsorten. *Z. Pflanzenzücht.* **30**, 292–299.

Scheffer, R. P. (1976). Host-specific toxins in relation to pathogenesis and disease resistance. *In* "Encyclopaedia of Plant Physiology, Vol. 4" (R. Heitefuss and P. H. Williams, eds.), pp. 247–269. Springer-Verlag, Berlin and New York.

Scheffer, R. P., and Yoder, O. C. (1972). Host-specific toxins and selective toxicity. *In* "Phytotoxins in Plant Diseases" (R. K. S. Wood, A. Ballio, and A. Graniti, eds.), pp. 251–272. Academic Press, New York.

Scheibe, A. (1930). Studien zum Weizenbraunrost, *Puccinia triticina* Erikss. II. Über die anfälligkeit von Weizensorten gegenüber verschiedenen Braunrost-Biotypen in den einzelnen Entwicklungsstadien der Wirtspflanzen. *Arb. Biol. Reichsanst. Land Fortwirtsch. Berlin-Dahlem* **17**, 549–585.

Scheifele, G. L., Nelson, R. R., and Koons, C. (1969). Male sterility cytoplasm conditioning susceptibility of resistant inbred lines of maize to yellow leaf blight caused by *Phyllosticta zeae*. *Plant Dis. Rep.* **53**, 656–659.

Schick, R. (1932). Über das Verhalten von *Solanum demissum*, *Solanum tuberosum* und ihren Bastarden gegenüber verschiedenen Herkünften von *Phytophthora infestans*. *Züchter* **4**, 233–237.

Schick, R., Möller, K. H., Haussdörfer, M., and Schick, E. (1958a). Die Widerstandsfahigkeit von Kartoffelsorten gegenüber der durch *Phytophthora infestans* (Mont.) de Bary hervorgerufenen Krautfäule. *Züchter* **28**, 99–105.

Schick, R., Schick, E., and Haussdorfer, M. (1958b). Ein Beitrag zur physiologischen Sperzialisierung von *Phytophthora infestans*. *Phytopathol. Z.* **31**, 225–236.

Schieber, E. (1975). Present status of coffee rust in South America. *Annu. Rev. Phytopathol.* **13**, 375–382.

Schnathorst, W. C. (1959). Resistance in lettuce to powdery mildew related to osmotic values. *Phytopathology* **49**, 562–571.

Shaner, G., and Hess, F. D. (1978). Equations for integrating components of slow leaf-rusting resistance in wheat. *Phytopathology* **68**, 1464–1469.

Shaner, G., Ohm, H. W., and Finney, R. E. (1978). Response of susceptible and slow-rusting wheats to infection by *Puccinia recondita*. *Phytopathology* **68**, 471–475.

Shattock, R. C. (1976). Variation in *Phytophthora infestans* on potatoes grown in walk-in polyethylene tunnels. *Ann. Appl. Biol.* **82**, 227–232.

Shepard, J. F., Bidney, D., and Shahin, E. (1980). Potato protoplasts in crop improvement. *Science* **208**, 17–24.

Sidhu, G. S. (1975). Gene-for-gene relationships in plant parasitic systems. *Sci. Prog.* **62**, 467–485.

Sidhu, G. S., and Webster, J. M. (1974). Genetics of resistance in the tomato to root-knot nematode: wilt-fungus complex. *J. Hered.* **65**, 153–156.

Sidhu, G. S., and Webster, J. M. (1979). Genetics of tomato resistance to the Fusarium: Verticillium complex. *Physiol. Plant Pathol.* **15**, 93–98.

Simons, M. D. (1969). Heritability of crown rust tolerance in oats. *Phytopathology* **59**, 1129–1133.

Simons, M. D. (1975). Heritability of field resistance to the oat crown rust fungus *Puccinia coronata*. *Phytopathology* **65**, 324–328.

Simons, M. D. (1979). Modifications of host-parasite interactions through artificial mutagenesis. *Annu. Rev. Phytopathol.* **17**, 75–96.

Simons, M. D., and Michel, L. J. (1968). Oat maturity and crown rust response. *Crop Sci.* **8**, 254–256.

Skovmand, B., Wilcoxson, R. D., Shearer, B. L., and Stucker, R. E. (1978). Inheritance of slow rusting to stem rust in wheat. *Euphytica* **27**, 95–107.

Smith, D. R. (1975). Expression of monogenic chlorotic lesion resistance to *Helminthosporium maydis* in corn. *Phytopathology* **65**, 1160–1165.

Stakman, E. C., Stewart, D. M., and Loegering, W. Q. (1962). Isolation of physiologic races of *Puccinia graminis* var. *tritici*. *U.S. Dep. Agric. Agric. Res. Serv. Bull.* **E617.**

Stall, R. E. (1961). Development of Fusarium wilt on resistant varieties of tomatoes caused by a strain different from race 1 isolates of *Fusarium oxysporum* f. *lycopersici*. *Plant Dis. Rep.* **45**, 12–15.

Stall, R. E., and Walter, J. M. (1965). Selection and inheritance of resistance in tomato to isolates of races 1 and 2 of *Fusarium* wilt organism. *Phytopathology* **55**, 1213–1215.

Statler, G. D., Nordgaard, J. T., and Watkins, J. E. (1977a). Slow leaf rust development on durum wheat. *Can. J. Bot.* **55**, 1539–1543.

Statler, G. D., Watkins, J. E., and Nordgaard, J. (1977b). General resistance displayed by three hard red spring wheat (*Triticum aestivum*) cultivars to leaf rust. *Phytopathology* **67**, 759–762.

Stevenson, F. T., Akeley, R. V., and Webb, R. E. (1955). Reactions of potato varieties to late blight and insect injury as reflected in yields and percentage solids. *Am. Potato J.* **32**, 215–221.

Sztejnberg, A., and Wahl, I. (1976). Mechanism of stability of slow stem rusting resistance in *Avena sterilis*. *Phytopathology* **66**, 74–80.

Todd, W. A., Green, R. J., and Horner, C. E. (1977). Registration of Murray Mitchum peppermint. *Crop Sci.* **17**, 188.

Tokeshi, H., Galli, F., and Kurozawa, C. (1966). Nova raça de Fusarium do tomateiro em São Paulo. *Anais Esc. Sup. Agric. "Luis Quiroz"* **23**, 217–227.

Tomerlin, J. R., Eversmeyer, M. G., Kramer, C. L., and Browder, L. E. (1983). Temperature and host effects on latent and infectious periods and on urediniospore production of *Puccinia recondita* f. sp. *tritici*. *Phytopathology* **73**, 414–419.

Toxopeus, H. J. (1956). Reflections on the origin of new physiological races of *Phytophthora infestans* and the breeding for resistance in potatoes. *Euphytica* **5**, 221–237.

Trelease, S. F., and Trelease, H. M. (1929). Susceptibility of wheat to mildew as influenced by carbohydrate supply. *Bull. Torrey Bot. Club* **56**, 65–92.

Vanderplank, J. E. (1963). "Plant Diseases: Epidemics and Control." Academic Press, New York.

Vanderplank, J. E. (1968). "Disease Resistance in Plants." Academic Press, New York.

Vanderplank, J. E. (1975). "Principles of Plant Infection." Academic Press, New York.

Vanderplank, J. E. (1978). "Genetic and Molecular Basis of Plant Pathogenesis." Springer-Verlag, Berlin and New York.

Vanderplank, J. E. (1982). "Host–Pathogen Interactions in Plant Disease." Academic Press, New York.

van der Zaag, D. E. (1956). Overwintering en epidemiologie van *Phytophthora infestans*, tevens enige nieuwe bestrijdingsmogelijkheden. *Tijdschr. Plantenziekten* **62,** 89–156.

van der Zaag, D. E. (1959). Some observations on breeding for resistance to *Phytophthora infestans. Eur. Potato J.* **2,** 278–286.

Vavilov, N. I. (1949). "The Origin, Variation, Immunity, and Breeding of Cultivated Plants" (Transl. from Russian by K. S. Chester). Chronica Botanica, Waltham, Massachusetts.

Verna, M. M., Kumar, J., and Kuchmar, S. (1975). Variability of horizontal resistance against brown rust (*Puccinia recondita* f. sp. *tritici*) in strains of *Triticum aestivum* and their progeny. *Cereal Res. Commun.* **3,** 149–162.

Vidhyasekaran, P. (1974). Possible role of sugars in restriction of lesion development in finger millet leaves infected with *Helminthosporium tetramera. Physiol. Plant Pathol.* **4,** 457–467.

Vohl, G. T. (1938). Untersuchungen über den Braunrost des Weizens. *Z. Pflanzenzücht.* **22,** 233–270.

Vowinkel, O. (1926). Die Anfälligkeit deutscher Kartoffelsorten gegenüber *Phytophthora infestans* (Mont.) de By. unter besonderer Berücksichtigung der Untersuchungsmethoden. *Arb. Biol. Reichsanst. Land Fortwirtsch. Berlin-Dahlem* **14,** 588–641.

Walker, J. C. (1966). The role of pest resistance in new varieties. *In* "Plant Breeding" (K. J. Frey, ed.), pp. 219–242. Iowa State Univ. Press, Ames.

Walter, J. M. (1967). Hereditary resistance to disease in tomato. *Annu. Rev. Phytopathol.* **5,** 131–162.

Waller, J. M. (1976). The influence of climate on the incidence and severity of some diseases of tropical crops. *Rev. Plant Pathol.* **55,** 185–194.

Ward, H. M. (1890). On the relations between host and parasite in certain epidemic diseases of plants. *Proc. R. Soc. London* **47,** 393–443.

Ward, H. M. (1902). On the relations between host and parasite in the bromes and their brown rust, *Puccinia dispersa* (Erikss.). *Ann. Bot.* **16,** 233–315.

Ward, H. M. (1904). On the history of *Uredo dispersa* (Erikss.) and the "mycoplasm" hypothesis. *Philos. Trans. R. Soc. London Ser. B* **196,** 29–46.

Watson, I. A., and Luig, N. H. (1963). The classification of *Puccinia graminis* var. *tritici* in relation to breeding resistant varieties. *Proc. Linn. Soc. N. S. W.* **88,** 235–258.

Watson, I. A., and Luig, N. H. (1968a). The ecology and genetics of host-pathogen relationships in wheat rusts in Australia. *Proc. 3rd Int. Wheat Genet. Symp.*, pp. 227–238.

Watson, I. A., and Luig, N. H. (1968b). Progressive increase of virulence in *Puccinia graminis* f. sp. *tritici. Phytopathology* **58,** 70–73.

Watson, I. A., and Singh, D. (1952). The future for rust resistant wheat in Australia. *J. Aust. Inst. Agric. Sci.* **18,** 190–197.

Webb, R. E., and Bonde, R. (1956). Physiological races of the late blight fungus from potato dump-heap plants in Maine in 1955. *Am. Potato J.* **33,** 53–55.

Weinhold, A. R., and English, H. E. (1964). Significance of morphological barriers and osmotic pressure in resistance of mature peach leaves to powdery mildew. *Phytopathology* **54,** 1409–1414.

Wellman, F. L. (1972). "Tropical American Plant Disease." Scarecrow Press, Metuchen, New Jersey.

Wolfe, M. S. (1978). Some practical implications of the use of cereal variety mixtures. *In* "Plant Disease Epidemiology" (P. R. Scott and A. Bainbridge, eds.), pp. 201–207. Blackwell, Oxford.

Wolfe, M. S. and Barrett, J. A. (1977). Population genetics of powdery mildew epidemics. *Ann. N. Y. Acad. Sci.* **287,** 151–163.

Wysong, D. S., and Hooker, A. L. (1966). Relation of soluble solids content and pith condition to *Diplodia* stalk rot in corn hybrids. *Phytopathology* **56,** 26–35.

Yarwood, C. E. (1934). The comparative behavior of four cloverleaf parasites on excised leaves. *Phytopathology* **24,** 797–806.

Yoder, O. C. (1980). Toxins in pathogenesis. *Annu. Rev. Phytopathol.* **18,** 103–129.

Yoder, O. C. (1981). Genetic analysis as a tool for determining the significance of host-specific toxins and other factors in disease. *In* "Plant Disease Control" (R. C. Staples and G. H. Toenniessen, eds.), pp. 3–11. Wiley, New York.

Zhukovsky, P. M. (1961). Grundlagen der Introduktion der Pflanzen und Resistenz gegen Krankheiten. *Züchter* **31,** 248–253.

Index

A

ABC–XYZ classification, *see* Virulence, reflected
Adaptation, of pathogen, *see also* Mutation
  direct, 40–43
  indirect, 43–47
Aecidium cantensis, 10
Aggressiveness of pathogen, 57–58, 60–63
Alfalfa, *see Colletotrichum trifolii; Uromyces striatus*
Alternaria solani, 2, 107
Amino acid residues, 100–105
Apple, *see Venturia inaequalis*
Area under disease progress curve, 145, 156

B

Barley, *see Erysiphe graminis hordei; Puccinia graminis; Puccinia hordei; Puccinia recondita; Puccinia striiformis; Ustilago hordei*
Barley yellow dwarf virus, 52
Biotrophy, 12, 101, 114–115, 117, 119
Bond
  hydrogen, 94
  hydrophobic, 94

C

Carbohydrate, 107–112, 119, *see also* Sugar
Cercospora sorghi, 20
Cochliobolus heterostrophus, see Helminthosporium maydis
Coffee, *see Hemileia vastatrix*
Colletotrichum atramentarium, 11
Colletotrichum circinans, 20
Colletotrichum dematium, 11
Colletotrichum graminicola, 11, 24, 108, 110
Colletotrichum lindemuthianum, 12
Colletotrichum trifolii, 11, 117
Colocasia esculenta, 162–163
Cotton, *see Glomerella gossypii; Phakopsora gossypii; Phymatotrichum omnivorum; Puccinia cacabata; Verticillium alboatrum; Xanthomonas malvacearum*
Cucumber mosaic, 22

D

Diagonal check, 5–7
Diplodia maydis, 108–111
Disease
  endemic, 162–168
  epidemic, 162–164, 166–167
Disease program, 131–132, 149–150

E

Eleusine corocana, 113
Epistasis, *see* Resistance; Variance; Virulence
Erwinia chrysanthemi zeae, 108
Erysiphe chicoracearum, 121
Erysiphe graminis hordei, 18, 67, 97, 118, 153, 172
Erysiphe graminis tritici, 54, 97, 118
Erysiphe polygoni, 117
Exserohilum turcicum, see Helminthosporium turcicum

F

Flax, *see Melampsora lini*
Fulvia fulva, 97
Fusarium moniliforme, 108
Fusarium moniliforme var. *subglutinans*, 108
Fusarium oxysporum f. sp. *conglutinans*, 79, 146
Fusarium oxysporum f. sp. *lycopersici*, 15–16, 76–79
Fusarium oxysporum f. sp. *niveum*, 1

G

Gene, *see also* Resistance; Virulence
    flow, 38
    number, 25
    pairs, diallel, 29–31
Gene-for-gene hypothesis, 3–7, 92–93, 96–98
Genetic engineering, 2–4
Gibberella zeae, 108
Globodera rostochiensis, 97
Glomerella glycines, 11
Glomerella gossypii, 11

H

Helminthosporium carbonum, 11
Helminthosporium maydis, 14, 109
Helminthosporium maydis race T, 11, 20–21, 109
Helminthosporium sorokinianum, 113–114
Helminthosporium tetramera, 113
Helminthosporium turcicum, 22–23, 26, 109–110
Helminthosporium victoriae, 11, 14, 172

Hemibiotrophy, 12, 101, 114, 117
Hemileia vastatrix, 68–69, 97, 167
Horizontal resistance equivalent, 83, 91
Hypersensitivity, 7

I

Interaction, 57–58, 60–61, 74–75, 77, 79–80
Intertropical convergence zone, 168

L

Latent period, 151–152, 156, 165–167
Logarithmic increase of disease, 133–136

M

Macrophomina phaseolina, 20, 108
Maize, *see Colletotrichum graminicola; Diplodia maydis; Erwinia chrysanthemi zeae; Fusarium moniliforme; Fusarium moniliforme* var. *subglutinans; Gibberella zeae; Helminthosporium maydis; Helminthosporium turcicum; Macrophomina phaseolina;* Maize dwarf mosaic virus; Maize stunt virus; *Phyllosticta zeae; Physopella zeae; Puccinia polysora; Puccinia sorghi; Pythium aphanidermatum*
Maize dwarf virus, 20
Maize stunt virus, 20, 23
Mayetiola destructor, 97
Melampsora lini, 18, 34, 49, 97
Meloidogyne incognita, 16
Mixture, varietal, 171–172
Multiline, 171–175
Mutation
    in crop plant, 2
    in pathogen, 49–56
    reciprocal in plant and pathogen, 106

N

Necrotrophy, 12, 108, 114–116, 117, 119

O

Oat, *see Puccinia coronata; Puccinia graminis avenae; Ustilago avenae*
Orobanche, 97

P

Peppermint, 2
Periconia circinata, 20, 79
Peronosclerospora sorghi, 20
Peroxidase, 102–105
Phakopsora gossypii, 10
Phakopsora pachyrhizi, 10
Phyllosticta zeae, 20
Phymatotrichum omnivorum, 5, 108
Physopella zeae, 9–10
Phytoalexin, 7
Phytophthora colocasiae, 162
Phytophthora infestans, 7–9, 12, 51, 53–54,
    58–61, 67–68, 70, 83–88, 90, 96–97, 100–
    101, 114, 122–129, 137–138, 147–153,
    162, see also Gene-for-gene hypothesis;
    Mutation; Vertifolia effect
Poa pratensis, 113–115
Potato, *see Aecidium cantensis; Alternaria
    solani; Colletotrichum atramentarium;
    Globodera rostochiensis; Phytophthora
    infestans; Potato virus* X; *Puccinia pit-
    tieriana; Streptomyces scabies; Synchy-
    trium endobioticum*
Potato virus X, 97
Progeny/parent ratio, 164–167, 169
Protein-for-protein hypothesis, 92–93
Protein polymerization, 104–105
Protein polymorphism, 92
Pseudomonas andropogoni, 20
Pseudomonas mors-prunorum, 53
Pseudomonas solanacearum, 79
Pseudomonogenes, *see* Resistance, pseudo-
    monogenic
Pseudospecificity, 79
Puccinia cacabata, 10
Puccinia coronata, 102, 118, 155, 159, 172
Puccinia graminis avenae, 94–95, 97
Puccinia graminis tritici, 9–10, 18–19, 24–25,
    28–33, 35–37, 39–50, 53–54, 56, 66, 71,
    88–90, 93, 97, 100–102, 128–129, 155,
    158, 172–173, 175, see also Adapta-
    tion of pathogen; Mutation; Resistance,
    ghost; Resistance, pseudomonogenic;
    Virulence, reflected, ABC–XYZ classifi-
    cation
Puccinia helianthi, 97
Puccinia hordei, 10, 145, 158
Puccinia pittieriana, 10

Puccinia polysora, 9–10, 102
Puccinia purpurea, 20
Puccinia recondita, 6, 9–10, 17, 71–74, 88, 97,
    100–102, 155, 158, 173, *see also* Resist-
    ance, slow-rusting
Puccinia sorghi, 9–10, 15, 23, 26, 97, 102, 118,
    156–158
Puccinia striiformis, 9–10, 14–15, 53, 61–63,
    97, 155, 173
Pyricularia oryzae, 65
Pythium aphanidermatum, 108

R

Race, definition, 37–38
Raspberry ringspot virus, 52, 79
Resistance, *see also* Gene-for-gene hypothesis;
    Temperature
  adult-plant, 72–74
  breeding for, 160–161, 167, 173–175
  continuously variable, 21–24
  cytoplasmic, 20–21
  discontinuously variable, 16–21
  ghost, 83, 88–90
  heritability, 155, 157
  horizontal, 57–61, 72, 74–75, 77–80, 116,
    129–131, 146–153
  late-rusting, 156
  ontogenic effect, 153, 159
  partial, 71–72, 95–96
  polygenic, 21–22, 24–25
  pseudomonogenic, 16, 18, 39, 48, 72, 122,
    143
  race-specific, unknown, 9
  race-unspecific, universal, 9
  remnant
    in horizontal resistance, 80
    in vertical resistance, 82–83
  sink-induced loss, 107–111, 159–160
  slow-rusting, 156
  unspecific, 5–12
  vertical, 47, 57–74, 95, 116, 122–130, 135–
    136, 139–142, 167, 171
Resistance gene numbers, 16–19, 93–94, 100–
    101
Resistance gene paradox, 7
Rhizobium, 97
Rice, *see Pyricularia oryzae; Xanthomonas
    oryzae*

S

Selection
  destabilizing, 29, 35–37
  stabilizing
    in horizontal resistance, 80–81
    in vertical resistance, 28, 55, 67, 175
Sorghum, 20
Soybean, *see Colletotrichum dematium;
  Glomerella glycines; Phakopsora pachy-
  rhizi*
Sphaerotheca pannosa, 120
Sphaerotheca reiliana, 20
Stemphylium sarcinaeforme, 117
Streptomyces scabies, 2
Sugar, 115–121
Sunflower, *see Orobanche; Puccinia helianthi*
Susceptibility
  endothermic, 94, 96
  specific, 5–12, 92–93
Synchytrium endobioticum, 69–70, 97

T

Temperature, 14–15, 19, 33, 94–96
Tilletia caries, 97
Tilletia contraversa, 97
Tilletia foetida, 97
Tobacco mosaic virus, in tomato, 15, 52, 97
Tomato spotted wilt virus, 51, 97
Toxin
  host-specific, 11
  osmotic, 117, 120–121
Transgression, 152, 157, 161

U

Uncinula necator, 121
Uromyces fallens, 117
Uromyces striatus, 11
Ustilago avenae, 97

Ustilago hordei, 97
Ustilago tritici, 97

V

Variable, third, *see* Interaction, higher-order
Variance
  additive, 14, 26–27, 155
  dominance, 14–15
  environmental, 14
  epistatic, 15, 33
  partitioning, difficulty in, 26–27
Vavilov's Rule, 100
Venturia inaequalis, 97, 101
Verticillium albo-atrum, 2, 15–16, 108
Vertifolia effect, 82–88, 113
Virulence, *see also* Interaction; Selection
  association, 29, 35–37, 44, 46
  definition, 57–58
  dissociation, 29, 31–33, 45
  reflected, ABC–XYZ classification, 29–37,
  44–48, 175

W

Watermelon, 1
Wheat, *see Colletotrichum graminicola; Ery-
  siphe graminis; Mayetiola destructor;
  Puccinia graminis; Puccinia recondita;
  Puccinia striiformis; Tilletia caries; Til-
  letia contraversa; Tilletia foetida; Usti-
  lago tritici*
Wounds, 102–103

X

Xanthomonas holcicola, 20
Xanthomonas malvacearum, 18, 52–53, 97,
  101, 114
Xanthomonas oryzae, 53